T0076440

AI FOR PHYSICS

Written in accessible language without mathematical formulas, this short book provides an overview of the wide and varied applications of artificial intelligence (AI) across the spectrum of physical sciences. Focusing in particular on AI's ability to extract patterns from data, known as machine learning (ML), the book includes a chapter on important machine learning algorithms and their respective applications in physics. It then explores the use of ML across a number of important sub-fields in more detail, ranging from particle, molecular, and condensed matter physics to astrophysics, cosmology, and a candidate for a theory of everything. The book covers such applications as the search for new particles and the detection of gravitational waves from the merging of black holes and concludes by discussing what the future may hold.

AI FOR EVERYTHING

Artificial intelligence (AI) is all around us. From driverless cars to game winning computers to fraud protection, AI is already involved in many aspects of life, and its impact will only continue to grow in future. Many of the world's most valuable companies are investing heavily in AI research and development, and not a day goes by without news of cutting-edge breakthroughs in AI and robotics.

The *AI for Everything* series will explore the role of AI in contemporary life, from cars and aircraft to medicine, education, fashion and beyond. Concise and accessible, each book is written by an expert in the field and will bring the study and reality of AI to a broad readership including interested professionals, students, researchers, and lay readers.

AI for Immunology
Louis J. Catania

AI for Cars
Josep Aulinas & Hanky Sjafrie

AI for Digital Warfare
Niklas Hageback & Daniel Hedblom

AI for Art
Niklas Hageback & Daniel Hedblom

AI for Creativity
Niklas Hageback

AI for Death and Dying
Maggi Savin-Baden

AI for Radiology
Oge Marques

AI for Games
Ian Millington

AI for School Teachers
Rose Luckin, Karine George & Mutlu Cukurova

AI for Learning
Carmel Kent & Benedict du Boulay

AI for Social Justice
Alan Dix and Clara Crivellaro

AI for the Sustainable Development Goals
Henrik Skaug Sætra

AI for Healthcare Robotics
Eduard Fosch-Villaronga & Hadassah Drukarch

AI for Physics
Volker Knecht

For more information about this series please visit:
https://www.routledge.com/AI-for-Everything/book-series/AIFE

AI FOR PHYSICS

VOLKER KNECHT

WITH CONTRIBUTIONS FROM

KILIAN HIKARU SCHEUTWINKEL
MARIO CAMPANELLI
MAYANK AGRAWAL
ÁLVARO DÍAZ FERNÁNDEZ
CHAO FANG
DANIEL GRÜN
BERNARD JONES
JIMENA GONZÁLEZ LOZANO
YANG-HUI HE

CRC Press is an imprint of the
Taylor & Francis Group, an **informa** business

First edition published 2023
by CRC Press
6000 Broken Sound Parkway NW, Suite 300, Boca Raton, FL 33487-2742

and by CRC Press
4 Park Square, Milton Park, Abingdon, Oxon, OX14 4RN

CRC Press is an imprint of Taylor & Francis Group, LLC

ISBN: 978-1-032-15655-2 (hbk)
ISBN: 978-1-032-15169-4 (pbk)
ISBN: 978-1-003-24518-6 (ebk)

DOI: 10.1201/9781003245186

Typeset in Joanna
by Deanta Global Publishing Services, Chennai, India

Access the Support Material: https://www.routledge.com/9781032156552

CONTENTS

Part II Machine-Learning the World from Subatomic to Cosmic Scales

ACKNOWLEDGMENTS

First, I (V. K.) want to thank various people without whom this book would never have been written.

First and foremost I thank all coauthors for joining the team, the good collaboration, and their excellent contributions.

Furthermore, I acknowledge the team of CRC Press/Taylor & Francis Group editors who worked so closely with me to make this important and timely text possible. I am grateful for the patience and caring of my acquisition editor, Elliott Morsia, who invited me to write this book, and the always helpful assistant editor, Talithah Duncan-Todd.

Likewise, I thank Assistant Professor Dr. Mathias S. Scheurer from the University of Innsbruck in Austria for his invariably positive review of my book proposal.

I also emphasize that this book could not have been created without the invaluable assistance of Google Scholar.

And finally, I give my deepest thanks and love to my wife Joschi Mansfield, for encouraging me to start writing about science for the general public, and for her continuous support and patience.

CONTRIBUTORS

Volker Knecht, Germany, Editor at *International Journal of Molecular Sciences*, freelance science writer, Diploma in Physics at University of Kaiserslautern, PhD in Theoretical Physics at University of Göttingen, PhD project at MPI Göttingen, postdoc at University of Groningen, group leader and PI at MPI Potsdam and University of Freiburg. Research at the interface between physics, chemistry, biology, and computer science for 17 years.

Mayank Agrawal is postdoc at Brown University School of Engineering and Research Affiliate at Massachusetts Institute of Technology (MIT), Massachusetts, USA. He earned his BSc and MSc degrees from the Indian Institute of Technology in Delhi and PhD in chemical engineering from the Georgia Institute of Technology. Expert in machine learning, computational chemistry, molecular simulation, nanoporous materials, adsorptive separation, and electrocatalysis. His current research includes combining ab-initio methods with machine learning for accelerated electrocatalysis simulations.

Mario Campanelli is Professor, Department of Physics and Astronomy, University College London, London, United Kingdom.

Over 25 years of experience on particle physics experiments like L3, ICARUS, HARP, CDF, ATLAS, DUNE. Expert in jet physics and forward detectors.

Álvaro Díaz Fernández is Assistant Professor at Universidad Politécnica de Madrid, Spain. He holds an MSc in condensed matter physics and biological systems from Universidad Autónoma de Madrid and a PhD in condensed matter physics from Universidad Complutense de Madrid. His research focuses on the theoretical understanding of the electronic properties of topological materials.

Chao Fang is Software Engineer at Google, Seattle, Washington, USA. He holds a BSc in astrophysics from Beijing Normal University, an MSc in theoretical and mathematical physics from the Chinese Academy of Sciences, and a PhD in computational physics (machine learning in condensed matter physics) from Texas A&M University.

Daniel Grün is Professor and Chair of Astrophysics, Cosmology, and Artificial Intelligence at the Physics Department and University Observatory of the Ludwig Maximilian University of Munich (LMU), Germany. He holds an MSc and a PhD from the Ludwig Maximilian University of Munich and went through training and career stages at Stanford University and SLAC National Accelerator Laboratory, as well as Stanford's Kavli Institute for Particle Astrophysics and Cosmology. His focus includes gravitational lensing, photometric surveys, and large-scale structure.

Yang-Hui He (BA Princeton, MSt Cambridge, PhD MIT) is Fellow of London Institute, Royal Institution; Tutor at Merton College, Oxford University; Professor of Mathematics at City, University of London; and Chair in Physics, Nankai University. He works on the interface between string theory, algebraic geometry, and machine learning.

Bernard Jones is a cosmologist and astrophysicist with training/ career stages/sabbatical visits in Cambridge (student and faculty member), Princeton, Berkeley, Paris, Copenhagen, Leiden, and London (Imperial College), and Emeritus Professor at Rijksuniversiteit Groningen, the Netherlands. Together with colleagues he set up the Theoretical Astrophysics Centre in Copenhagen.

Jimena González Lozano is a physics PhD student at the University of Wisconsin–Madison, USA. She is working in observational cosmology and astrophysics and is interested in machine learning and data science.

Kilian Hikaru Scheutwinkel is a physics PhD candidate in astrophysics at the University of Cambridge. He holds an MSc in Physics from Uppsala University. Furthermore, he was a Senior Consultant and Software Engineer in Tokyo and Research Assistant at universities in Stockholm and Frankfurt am Main. He uses Bayesian machine learning approaches to analyze Big Data generated by the radio telescopes REACH and HERA.

LIST OF ABBREVIATIONS

AI	Artificial intelligence
CNN	Convolutional neural network
EDGES	Experiment to Detect the Global EoR Signature
FFT	Fast Fourier transform
GAN	Generative adversarial network
GD	Gradient descent
GW	Gravitational waves
KNN	k-nearest neighbors
ΛCDM	Lambda cold dark matter
LHC	Large Hadron Collider
LIGO	Laser Interferometer Gravitational-Wave Observatory
MCMC	Markov Chain Monte Carlo
ML	Machine learning
NN	Neural network
PCA	Principal component analysis
QFT	Quantum field theory
RBM	Restricted Boltzmann machine
ROC	Receiving operating characteristic
SGD	Stochastic gradient descent
SKA	Square Kilometer Array
SM	Standard Model (of particle physics)

SR Symbolic regression
TICA Time-lagged independent component analysis
TPC Time projection chamber
ToE Theory of everything

PART I

OPENING

1

GATHERING THE TEAM

VOLKER KNECHT

This short book is addressed to lay readers and gives a concise overview on the interplay between artificial intelligence (AI) and physics grasped in an afternoon read. Unlike textbooks for students, this treatise dispenses with mathematical formulas and instead presents the subject matter in an intuitive manner. Nevertheless, with its broad scope in compact format this work will also be interesting for professionals, students, and researchers.

How do AI and physics connect? Before we get into this, we will start our journey taking a global view on each field, AI and physics, separately. After emphasizing the historical evolution of both fields, we will pick up those threads later.

AI AND MACHINE LEARNING

AI refers to computer programs smartly performing tasks on their own, mimicking human intelligence. The first work now generally recognized as AI is McCullouch and Pitts' formal design for programmable "artificial neurons" published in 1943.[1] The term "artificial intelligence" was invented at a workshop at Dartmouth College in 1956, often considered as the birth of the field of AI.

DOI: 10.1201/9781003245186-2

Three years later, a pioneer of this field invented the term "machine learning" (ML). ML is a generic term for the "artificial" generation of knowledge from experience: a synthetic system learns from examples and can generalize these after the learning phase is complete. To do so, an ML algorithm builds a statistical model based on training data. This means it does not simply learn the examples by heart but recognizes patterns and regularities in the learning data. In this way, the system can also assess unknown data, though it may also fail to do so.

As detailed in Chapter 2, ML comprises many different tools. These include versions of search and mathematical optimization, methods based on statistics and probability, along with bio-inspired artificial neural networks, also simply called neural networks (NNs).

A NN contains neurons connected in a pair-wise manner and arranged in multiple tiers, comprising obligatory input and output layers, along with one or multiple optional hidden layers. The family of methods with or without hidden layers is denoted as deep or shallow learning, respectively. Deep learning is extremely powerful, much more so than shallow learning. It allows the approximation of a larger class of mappings of input to output data. While the first deep learning systems were developed in the 1960s, the term "deep learning" itself was introduced in 1986 to denote a procedure recording all used outcomes of a considered search space not having led to a desired solution.[2]

In the years since its birth in 1956, AI has experienced several *waves of optimism and pessimism*, the latter including two major AI winters in 1974–80 and 1987–93. After AlphaGo created by DeepMind* effectively beat a professional Go player in 2015, AI once again attracted extensive global attention. Since then, the complexity of top AI applications has been increased exponentially; the performance of ML setups doubles every three to four months.[3]

In 2020, AlphaFold by DeepMind made a breakthrough in protein folding, a "holy grail" problem in the field of biology. Here

---

* Formerly Google DeepMind, a company specialising in AI programming.

the task is to predict protein structure from the sequence of amino acids. AlphaFold achieved an accuracy regarded as comparable with experimental techniques.[4, 5] Providing this software, AI thus solved one of the hardest problems in science.

Besides protein folding and automatic game playing, AI is now increasingly used for a wide range of diverse applications. These include speech recognition, automatic language translation, internet search engines, product recommendations, and email spam filtering. Other purposes are stock market trading, image recognition, self-driving cars, medical diagnosis, drug discovery, and genomics.

How about physics? What can AI do for it? And what can physics do for AI? To judge this, we should first take a step back and look at physics by itself taking a global and historical perspective.

A BRIEF HISTORY OF PHYSICS

Physics is the natural science studying matter, its motion and behavior through space and time, as well as the related entities of energy and force. Compared to AI, physics has a much longer tradition. Through its inclusion of astronomy, physics is in fact the oldest science overall. The oldest known star chart (resembling the constellation Orion) is believed to be contained in a 32,500-year-old carved ivory mammoth tusk.[6] The first star catalogue is a Babylonian one and consists of the Three Stars Each texts whose appearance in early Mesopotamia dates back to the 12th century BC.[7]

Early astronomic studies consisted of the observation and predictions of the motions of objects visible to the naked eye. A rational, physical explanation for celestial phenomena and nature overall beyond pure observation was started in ancient Greece, initiated by the thinking of pre-Socratic philosophers during the Archaic period (650–480 BC). Thales of Miletus (7th and 6th centuries BC) is nicknamed "the Father of Physics" or "the Father of Science" for refusing to accept various supernatural, religious, or mythological explanations for natural phenomena, and proclaiming every event had a natural cause.[8] Leucippus (5th century BC) was the first to propose the physical world is composed of fundamental indivisible components known as atoms.* His pupils Democritus (460–370 BC) and Anaxagoras (499–428 BC) are said to have believed the Milky Way is a cluster of stars. This was confirmed much later after the invention of the telescope by Galileo Galilei in the 17th century AD.

From the 5th century BC in Classical Greece, it was common knowledge that the Earth is spherical ("round"). Aristarchus of Samos (310–230 BC) assessed the relative distances of the Sun and the Moon; he also proposed a model of the Solar System where the Earth and planets rotated around the Sun, now called the heliocentric model. As a tool to calculate the location of the Sun, Moon, and planets for a given date, Greek astronomers developed the Antikythera mechanism (c. 150–80 BC). This mechanic construct is generally referred to as the first known analog computer.

Important physical and mathematical traditions also occurred in ancient Chinese and Indian sciences (between 400 BC and the 15th century AD). In the 7th to 15th centuries AD, scientific progress appeared in the Muslim world.

During the 16th and 17th centuries, Europe was the stage for a scientific revolution. Nicolaus Copernicus (1473–1543) re-discovered the heliocentric model of the Solar System. His perspective, along with accurate observations made by Tycho Brahe, enabled Johannes Kepler (1571–1630) to articulate his laws regarding planetary motion still in use today. Supporting Copernicanism, Galileo Galilei pioneered the telescope and made ground-breaking astronomical discoveries along with empirical experiments, rendering him "the father of modern science".[9] Isaac Newton (1642–1727) combined his own discoveries in mechanics and astronomy with earlier ones to create a single system for describing the mechanisms of the universe, providing the fundamental laws of classical mechanics and a theory for gravitation.

In the 19th century, thermodynamics was developed, a branch of physics dealing with heat, work, and temperature, and their relation

---

* From the ancient Greek word "atomos" for "uncuttable".

to energy, radiation, and physical properties of matter. Experimental achievements by Alessandro Volta, André-Marie Ampère, and Michael Faraday along with theoretical work by James Clerk Maxwell led to a comprehensive understanding of electromagnetism. This culminated in the finding that light is an electromagnetic wave.

Lord Kelvin (1824–1907) and Rudolf Clausius (1822–88) formulated the law of conservation of energy. It is equivalent to the first law of thermodynamics stating the following: the change in the internal energy of a closed system is the heat supplied to the system minus the thermodynamic work done by the system on its surroundings. Kelvin and Clausius also stated the second law of thermodynamics implying that entropy (a measure of disorder) cannot decrease over time. Processes reversible in time (such as fluid flows in a well-designed turbine) are characterized by a constant entropy. In contrast, time-irreversible processes (such as the mixing of gases or fluids) occur along with an increase in entropy. The latter is often referred to as the concept of the arrow of time. Explaining this notion, as well as heat and in general the macroscopic behavior of nature (gases, fluids, or solids) from the behavior of large assemblies of microscopic entities (such as atoms or molecules), Ludwig Boltzmann (1844–1906) and Josiah Willard Gibbs (1839–1903) developed statistical mechanics, as a fundamentally new approach to science.[10]

In the 20th century, the physical view of the world was turned upside down by two revolutionary theories supported by a wealth of experimental data. The first one is Albert Einstein's theory of relativity transforming our perception of space, time, and matter. Special relativity says the velocity of light in vacuum is the ultimate speed limit in nature and a fundamental constant connecting space and time to a four-dimensional spacetime. General relativity (often termed the most beautiful of all existing physical theories) declares gravity as *curvature* of spacetime. The second theory is quantum mechanics; it says energy (like light) is transmitted in quanta and microscopic objects have both particle and wave character, with often incredibly bizarre consequences.

Physicists started to seriously investigate the world on extreme spatial scales, exceedingly small and large ones, with two distinct theoretical frameworks. On the one hand, there is cosmology. Its theoretical basis is general relativity, and it is concerned with studies of the origin and evolution of the universe, from the Big Bang to today and on into the future.

Research on extremely small scales focuses on subatomic particles, especially to understand the fundamental forces keeping the world together at its core. The corresponding theoretical framework is based on quantum mechanics and special relativity. From the 1930s, experimental studies of subatomic scales have taken particles apart by smashing them into each other using particle accelerators. As the small parts are held together by exceedingly large forces, high energies are required to break them apart, a major experimental challenge in this area.

In physical research overall, experimental tools probe natural phenomena explained and predicted by theory. The latter describes physical objects and systems using mathematical models. Traditionally, mathematical calculations have been done by hand. A new development was initiated by the first large-scale deployment of electronic computers in the Manhattan Project to mathematically model nuclear detonation. Born was the new technique of computer simulations, developing with the rapid growth of the computers since then.

Overall, physics is the basis for chemistry, biology, geology, and ecology and thus may be considered the most fundamental branch of natural sciences. Furthermore, it provides the foundations of engineering and thus for all modern technology.

REFERENCES

1. McCulloch, W. S.; Pitts, W., A logical calculus of the ideas immanent in nervous activity. *The Bulletin of Mathematical Biophysics* 1943, 5 (4), 115–133.
2. Dechter, R., *Learning while searching in constraint-satisfaction problems*. AAAI-86 Proceedings, 1986.

3. Saran, C., *Stanford University finds that AI is outpacing Moore's Law*. 12 Dec 2019 9:56 ed.; Saran, C., Ed. 2019; computerweekly.com.
4. Callaway, E., 'It will change everything': DeepMind's AI makes gigantic leap in solving protein structures. *Nature* 2020, 588 (7837), 203+.
5. Singh, A., Deep learning 3D structures. *Nature Methods* 2020, 17 (3), 249–249.
6. Whitehouse, D. Oldest star chart found 2003. http://news.bbc.co.uk/.
7. North, J., *The Norton history of astronomy and cosmology*. W. W. Norton & Company: 1995.
8. Thales Of Miletus: The father of physics. *History of Physics* [Online], 2020. http://www.history-of-physics.com/ (accessed May 21, 2021).
9. Disraeli, I., *Curiosities of literature*. Baudry's European Library: 1835; Vol. 1.
10. Gibbs, J. W., *Elementary principles in statistical mechanics*. Courier Corporation: 2014.

2

TEAMPLAY

VOLKER KNECHT

What do AI and physics have to do with each other? Well, they mutually affect each other in both directions, as discussed in the subsequent text.

MACHINE LEARNING PHYSICS

Experimental measurements produce more and more data requiring analysis. This is because science and technology are continuously improving in building experimental devices, such as accelerators to smash subatomic particles at ever-higher energies and increasing numbers of ever better telescopes to scan the sky. Citizen Science projects allow thousands of volunteers to be organized to classify astronomical objects by hand. For example, Galaxy Zoo has mobilized (as of April 21, 2021) almost 66,000 people to characterize more than 256,000 galaxies in terms of their shape.[1]

Manual classification, though, becomes difficult when the number of astronomic objects considered gets too large. In 2020, Clarke and colleagues used ML to classify 111 million light sources.[2] This is just a fraction of the data expected to be gathered very soon by next generation telescopes. The Large Synoptic Survey Telescope whose

DOI: 10.1201/9781003245186-3

commissioning is planned for October 2023 is expected to catalog approximately 20 billion (2×10^{10}) galaxies and a similar number of stars.[3] The total number of astronomic light sources awaiting the discovery is even larger; a 2016 study published in *The Astrophysical Journal* and led by Christopher Conselice of the University of Nottingham used 20 years of images from the Hubble Space Telescope to estimate that the observable universe contains at least two trillion (2×10^{12}) galaxies.[4] This demonstrates a huge need to use ML to classify astronomic objects.

A major challenge in handling physical data overall is to dissect tiny amounts of signal from overwhelming noise, like finding the needle in a haystack. This can be illustrated by an example from physics at the smallest scales studied using particle accelerators. These experimental tools are some of the most powerful ones available, propelling particles to nearly the speed of light and then colliding them to study the resulting interactions and particles forming.

The most powerful particle accelerator in the world is the Large Hadron Collider (LHC) in Geneva, Switzerland. Correspondingly impressive is its production: the LHC generates one petabyte of data per second.[5] This is far more than can be collected by even the world's leading research institutions.[6] Hence, it is crucial to quickly pick and choose the interesting events to keep on the fly, while getting rid of the rest. This is where ML comes into play.

Specifically, it was the LHC where the "God particle" was detected in 2012, as the last particle of the Standard Model of particle physics (summarized in Chapter 3).[7] This particle known as the Higgs boson was predicted in 1964 by several scientists including François Englert and Peter Higgs who both won the Nobel Prize in 2013. The Higgs boson is an elementary particle produced by the quantum excitation of the Higgs field. This field is essential as it gives mass to all fundamental constituents of matter (including electrons and quarks), except to itself.

Experimentally, the Higgs boson can be created by smashing protons into each other. A Higgs boson is produced every few billion proton–proton collisions. Like most of the particles generated in the LHC experiments, it decays before it can be detected by any sensor. Its cleanest experimental signature is its decay into muon–antimuon pairs occurring once every 10^{13} proton–proton collisions.[8]

The search for the Higgs boson required the most expensive experiment ever. It produced an enormous amount of data in which signatures of the Higgs particle were lost like the needle in a haystack. "How do you find the needle?", Tony Stark alias Iron Man is asked in the Marvel movie *Avengers: Age of Ultron*. He replies: "I use a magnet!" The "magnet" in the Higgs case is machine learning which helped to distinguish the Higgs signal from various non-Higgs backgrounds.[9]

Besides its use in analyzing observational data in experimental physics, ML may also be utilized in theoretical physics, as in the form of computer simulations. For instance, Pang and colleagues[10] applied computer simulations in conjunction with deep learning to study the phase transition of quarks and gluons, the fundamental particles making up protons and neutrons, the building blocks of atomic nuclei. At temperatures below the Hagedorn temperature of approximately 2 terakelvin, quarks are never isolated, denoted as *confinement*, and corresponding to ordinary matter. At higher temperatures, quarks and gluons form a plasma containing quarks as quasi-free particles. This phase is the quark–gluon plasma, the presumed state of matter in the time interval of 10^{-10}–10^{-6} s after the Big Bang.[11] Wang and colleagues used their simulations to relate the phase transition of quarks and gluons to the final-state spectra of heavy-ion collisions using calculations based on the theory of liquids in motion, called hydrodynamics.

Computer simulations with or without the help of ML are based on numerically solving mathematical equations found previously. However, ML may be even used to infer such equations themselves from raw data. Here, the task is to find a symbolic expression matching data from an unknown function, denoted as symbolic regression.

Traditionally, this task is fulfilled by humans often considered geniuses. For example, using the data tables of planetary orbits measured by Tycho Brahe, Johannes Kepler needed 4 years and 40 failed attempts to discover Mars' orbit was an ellipse. In a work published in 2020, Udrescu and Tekmark demonstrate such challenges can be solved much more quickly by a computer.[12]

Their ansatz was that, although this problem is likely to be extremely complex in principle, functions of practical interest often exhibit symmetries and other simplifying properties. The authors developed an ML algorithm combining a neural network with a suite of physics-inspired techniques. They applied it to 100 equations of the *Feynman Lectures on Physics*. These included Kepler's ellipse equation, Newton's law of gravity, several equations of statistical physics, and six equations from Einstein's theory of special relativity. All equations were (re-)discovered within less than 2 hours, most equations even within 20 seconds. This work demonstrates: symbolic regression could enable computers to discover useful and hitherto unknown physics in the future.

Thus, ML can be used to ultimately uncover physical principles and recognize significant variables. This is also suggested from work by Seif, Hafezi, and Jarzynski showing an ML algorithm can learn to discern the direction of time's arrow when provided with a system's microscopic trajectory as input. It even identifies work along with entropy production as the relevant observables.[13] In this sense, their study is a step toward the AI-driven discovery of physical concepts.

IMPACT OF PHYSICS ON MACHINE LEARNING

What does physics inversely do for AI? Well, it does two things: On the one hand, it drives the development of new hardware platforms involving quantum computers and/or analog computers. On the other hand, it can help understanding the software via providing a theoretical framework that enables a fundamental assessment of ML algorithms. Let us start with the latter.

STATISTICAL PHYSICS OF ML

ML yields a theoretical framework for understanding AI through the connection between statistical mechanics and learning theory. For example, it is not fully understood why deep learning works so well. This is often quoted. But what does it mean?

In 2017, Lin, Tekmark, and Rolnick formulated the issue in a mathematical manner and solved it from a physics perspective using statistical mechanics.[14] Most applications of deep learning are centered around approximating mappings between different sets of data (functions) containing many variables. The question is how can neural networks approximate functions well in practice, when the set of possible functions is exponentially larger than the set of practically possible networks?

Suppose we wish to classify megapixel gray-scale images into cats or dogs. If each pixel can take 1 of 256 values, there are $256^{1,000,000}$ possible images. For each one we wish to calculate the probability it portrays a cat. In this case, an arbitrary function is defined by a list of $256^{1,000,000}$ probabilities. This is way more numbers than there are atoms in the observable universe (10^{78} –10^{82}).

Lin, Tekmark, and Rolnick revealed the classes of probability distributions used in physics and ML. Thereby they showed neural networks perform a combinatorial "swindle", replacing exponentiation by multiplication. If there are $n = 10^6$ input variables and $\nu = 256$ possible values for each of them, the number of parameters is reduced from ν^n to $\nu \times n$ times some constant factor.

The accomplishment of this "fraud" is observed to depend fundamentally on physics: neural networks only work well for an exponentially tiny fraction of all possible input data. This sounds like really bad news. However, here is the rescue: due to the laws of physics the data sets considered in ML are also drawn from an exponentially tiny fraction of all imaginable data sets. What luck!

Statistical mechanics thus provides the theoretical framework enabling a fundamental perception of ML algorithms. Physics does not only help in comprehending the software, though. It also drives the development of new hardware platforms providing help with expensive information processing pipelines. How can this be accomplished? Well, keywords are quantum computing and analog computing.

ANALOG COMPUTERS

Today's computers usually function on a digital basis. In these devices, both the representation of numbers and the time evolution are discrete. Numbers are (mostly) represented in binary form, using only two digits (0 and 1), and processed in discrete time steps.

This contrasts with analog computers where both the calculated values and the time evolution are continuous. Analog computing was the prevalent form of high-performance computing well into the 1970s. Such computers, however, are hard to program. Furthermore, they are intrinsically noisy. Data represented by analog signals can easily be corrupted. For these reasons, analog computers were replaced by digital ones.

High precision, though, is not needed for some ML tasks. In general, many modern applications including pattern recognition and computer vision can yield results with appropriate quality even if many individual computation steps are executed imprecisely. Furthermore, improved levels of precision may be achieved by analog computing architectures via repeating operations and averaging the result, reducing the impact of noise.[15]

Programmability, on the other hand, is not required for special purpose computers dedicated to ML. An abstract mathematical object like a matrix may be directly represented as a two-dimensional physical array with the same number of rows and columns. Additionally, mathematical operations can be adopted by electric circuits. For example, multiplication may be implemented by an amplifier.[16]

Research on new hardware platforms for analog computers is an intriguing task for physics labs. Such machines may be based on electronics like digital computers. Alternatively, analog computing could be implemented via electrochemistry or optics. Interestingly, some of the standard building blocks in optics labs are remarkably like the way information is processed with neural networks.[17–20]

An analog computer may be made specifically for a certain class of tasks with its architecture corresponding to a particular physical process. Thus, the device can be both faster and more energy efficient than a digital computer, the improvement increasing with the complexity of the application. Concretely, Li and colleagues demonstrated an analog computer can be up to 200 times faster and up to 600 times more power efficient than a digital one.[21]

QUANTUM COMPUTERS

A new era of information processing may be entered via quantum computing. As the name suggests this technology is based on the laws of quantum mechanics. Thus, it works based on quantum-mechanical states processed according to quantum-mechanical principles.

The basic information unit of a classical computer is a bit adopting the values 0 and 1. Also a quantum computer encodes information in binary form. Instead of bits, though, it deals with quantum bits – qubits.

A qubit is a two-state quantum-mechanical system. An example is the spin of the electron whose two states can be taken as spin up and spin down. Contrasting classical physics, though, quantum mechanics allows a qubit to be a superposition of both states simultaneously. Through superposition, elementary particles can also be in more than one place at any given time. Second, an individual particle, such as an electron, can cross its own trajectory and interfere with the direction of its path; this so-called quantum interference is evident from double slit experiments at extremely low particle densities. Third, a group of particles can be prepared such that the quantum state of each particle of the group cannot be described independently of the state of the others; this phenomenon is denoted as quantum entanglement and may occur even when the particles are separated by a large distance (one of the most bizarre consequences of quantum physics Einstein called "spooky action at a distance").

As a result of these quantum effects, the computing power of quantum computers can increase exponentially with the number of qubits. This applies to various algorithms, including quantum neural networks; these artificial NN models are based on the principles of quantum mechanics.

Quantum computers can be implemented in various ways. A promising one used commercially* is to employ ion traps. Here, individual charged atoms (ions) are threaded together like a string of pearls by electromagnetic fields in a vacuum. The qubits are formed by two long-lived electronic states of the individual ions, for instance two hyperfine levels of the ground state. The number of qubits corresponds to the number of ions in the trap. The qubits are controlled by lasers interacting with the individual ions. Via moving the ions in the trap, the qubits can be coupled with each other and thus entangled.

The first quantum computer with two qubits was realized in 1998.[22] The number of qubits implemented increased steadily over time, from 7 qubits at the Los Alamos Laboratory in 2001 via over 50 qubits at IBM in 2017 to 72 qubits with Google in 2018. In 2020, a Chinese group claimed to have released a prototype quantum computer comprising 76 qubits able to calculate 100,000 billion times faster than today's best supercomputers.[23]

This is breath-taking! It must be noted, though, that their benchmark was tailored to a specific problem, and the poor performance of classical computers claimed by the Chinese group in comparison was doubted by scientists at Google.

In any case, quantum computing may enhance ML applications not only in terms of speed but can lead to entirely new algorithms (denoted as quantum machine learning).[24, 25]

---

* By the first publicly traded pure-play quantum computing company IonQ founded in 2015.

MACHINE LEARNING THE PHYSICAL WORLD FROM SUBATOMIC TO COSMIC SCALES

ML is widely used as a tool in physical sciences. A literature search on Google Scholar for research articles containing the word group "machine learning" in journals with "physics" in their names yields 19,000 hits overall (as of April 14, 2021). Among them, 14,100 (74 percent) are from the past 5 years. This indicates the exponentially growing importance of ML in physics lately. As detailed in the Supplement, the proportion of physics studies using AI steadily increased from below 1 percent between 2000 and 2005 to 19 percent in 2021 (when its yearly rise jumped from 3 percent in the previous year to 8 percent).

These data also show that cases presented here can only be a small representative selection, spread over important domains of physics, emphasizing hot research topics. Going beyond the examples above and to further illustrate how AI boosts fundamental science, most of the remainder of this book will highlight how ML has been recently used to address extremely difficult questions in physics at all scales. An outline of the content of these chapters is given in the following.

After introducing important ML algorithms in Chapter 3, Chapter 4 will discuss how ML is utilized in research on the nature of particles constituting matter and radiation. This refers to the smallest detectable particles (such as quarks and electrons) and the fundamental interactions required to explain their behavior. These include the electromagnetic field and the Higgs field mentioned above. You will be introduced to the Standard Model (SM) of particle physics, its limitations, and theories beyond the SM. We will then discuss how ML is applied in probing the SM at extremely high energies and searching for new physics beyond the SM. This involves primarily applying ML to analyzing the wealth of products of particle collisions, as indicated in Figure 2.1A.[8]

Quark triplets make up protons and neutrons constituting the nuclei of atoms, whose shells consist of electrons. Via sharing electrons of their outer shells, atoms may be held together in a pairwise

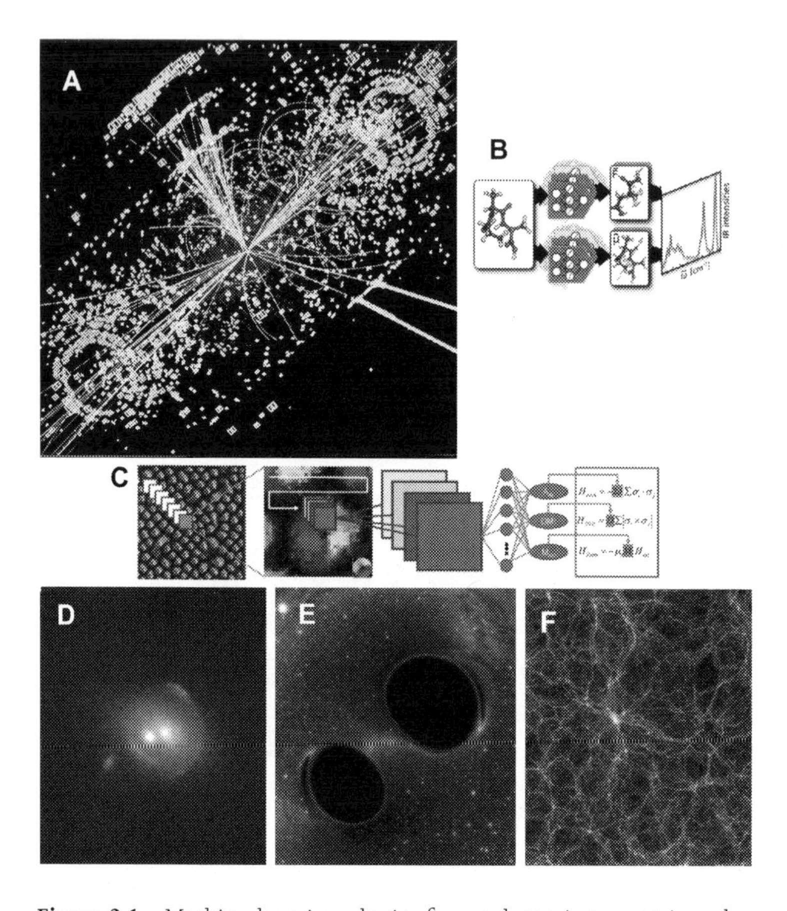

Figure 2.1 Machine learning physics from subatomic to cosmic scales. A: analyzing reaction products from elementary particle collisions (from Wikimedia Commons, by Lucas Taylor/CERN, File: CMS Higgs-event.j pg, under Creative Commons). B: computation of infrared spectra from molecular dynamics simulations (reproduced by permission of the Royal Society of Chemistry[26]). C: determination of magnetic properties of solid matter (reproduced by permission of Wiley[27]). D: gravitational lensing (by NASA, ESA/Hubble, and F. Courbin under CC). E: detecting gravitational waves from the merger of two black holes (from Wikimedia Commons, by Simulating eXtreme Spacetimes [SXS] project). F: simulating the cosmic web (from Wikimedia Commons, by unknown author).

manner in thousands, leading to tens of millions of different (macro) molecules. Research on molecular systems by computer simulations in conjunction with ML – as indicated in Figure 2.1B – will be addressed in Chapter 5.[26]

Many atoms or molecules of the same sort may arrange in various distinct states of matter called phases. If their mutual interactions are negligible, they form a gas. On the other hand, electromagnetic forces between them drive the formation of condensed phases. The most well-known ones are liquids and solids. Others are the ferromagnetic and antiferromagnetic phases of spins on crystal lattices of atoms. ML may be applied to such phases to infer magnetic properties, as illustrated in Figure 2.1C.[27] How ML is used to study condensed phases and transitions between them is described in Chapter 6.

The atoms and molecules in condensed phases are held together via electromagnetic forces. In contrast, gravity causes all bodies to fall downwards on earth and determines the orbits of planets, moons, and satellites, along with comets in the Solar System; in the cosmos, it determines the formation of stars and galaxies as well as its evolution on a large scale. The latter is the concern of cosmology. The current Standard Model of Cosmology assumes the universe was created in the "Big Bang" from pure energy and has been expanding since then, the latter being evident from the redshifts of the light from distant galaxies. As detailed in Chapter 7, ML is employed to detect and study processes and objects predicted from general relativity such as gravitational lensing[28] (shown in Figure 2.1D) or gravitational waves from the merger of black holes[29] (illustrated in Figure 2.1E), or to study the cosmic web (depicted in Figure 2.1F).[30]

Gravity is described by general relativity. The other three fundamental forces (the electromagnetic plus the strong and weak nuclear ones) are explained by the SM of particle physics. The SM and general relativity are fundamental as in principle the whole of established physics can be derived from them. However, these two pillars of physical science are not compatible with each other. This is because the SM is based on quantum mechanics which is incompatible with

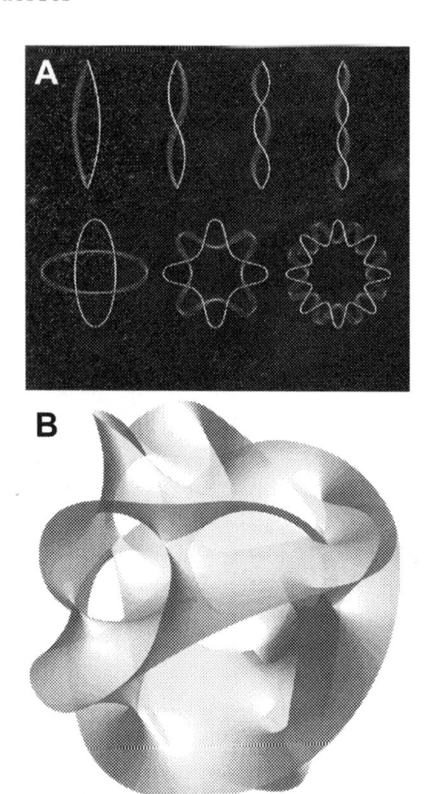

Figure 2.2 Machine learning theory of everything. A: strings, according to string theory the most fundamental objects of the universe (reprinted by permission from Springer Nature, Journal of Physics: Conference Series, IOP Publishing; Physics with Large Extra Dimensions and Non-Newtonian Gravity at Sub-mm Distances, Ignatios Antoniadis, copyright 2005[35]). B: compactification of extra dimensions (one out of incredibly many possible types; source: Wikimedia Commons). String theory is considered the most data-intensive study subject in science, thus crying out for the use of machine learning methods.

general relativity. What is missing is a quantum description of gravity. Quantum gravity is the Holy Grail of modern theoretical physics and (next to consciousness[31]) one of the two hardest problems in science today.[32] A promising candidate for an elegant theory of everything is string theory. Here the point-like particles of the SM (considered as coarse-grained descriptions) are replaced by one-dimensional objects called strings as the most fundamental objects (see Figure 2.2A).[33] Their vibration modes determine the properties of the elementary particles of the SM. Space is predicted to exhibit six extra dimensions compactified and thus not visible on a macroscopic scale. The detailed form of compactification determines the vibration modes of the strings. However, the set of possible choices of parameters governing the compactifications (denoted as string landscape, see Figure 2.2B) is incredibly large. Hence, string theory involves data considered the most comprehensive in science,[34] just crying out for the use of ML methods. How ML is used to study the string landscape is described in Chapter 8.

Finally, Chapter 9 gives a conclusion.

Before we highlight how the application of AI advances research in physics at all spatial scales, the following chapter will focus more closely on the methodologies provided by machine learning. It will detail artificial neural networks and other common ML algorithms.

REFERENCES

1. Galaxy Zoo. https://www.zooniverse.org/projects/zookeeper/galaxy-zoo/.
2. Clarke, A.; Scaife, A.; Greenhalgh, R.; Griguta, V., Identifying galaxies, quasars, and stars with machine learning: A new catalogue of classifications for 111 million SDSS sources without spectra. *Astronomy & Astrophysics* 2020, 639, A84.
3. Ivezić, Ž.; Kahn, S. M.; Tyson, J. A.; Abel, B.; Acosta, E.; Allsman, R.; Alonso, D.; AlSayyad, Y.; Anderson, S. F.; Andrew, J., LSST: From science drivers to reference design and anticipated data products. *The Astrophysical Journal* 2019, 873 (2), 111.

4. Brooke, L. There are at least two trillion galaxies in the universe, ten times more than previously thought. https://www.nottingham.ac.uk/ (accessed May 20, 2021).
5. Gaillard, M. CERN Data Centre passes the 200-petabyte milestone. *CERN Accelerating Science* [Online], 2017. https://home.cern/news/news/computing.
6. Combining big data analytics and deep learning for the large hadron collider. https://openlab.cern/sites/default/files/2019-09/Combining%20Big%20Data%20Analytics%20and%20Deep%20Learning%20for%20the%20Large%20Hadron%20Collider%20-%20Intel%20AI.pdf (accessed May 27, 2021).
7. Brown, A. The "God Particle". *Advanced Science News* [Online], 2019. https://www.advancedsciencenews.com/.
8. Radovic, A.; Williams, M.; Rousseau, D.; Kagan, M.; Bonacorsi, D.; Himmel, A.; Aurisano, A.; Terao, K.; Wongjirad, T., Machine learning at the energy and intensity frontiers of particle physics. *Nature* 2018, 560 (7716), 41–48.
9. Bourilkov, D., Machine and deep learning applications in particle physics. *International Journal of Modern Physics A* 2019, 34 (35), 1930019.
10. Pang, L.-G.; Zhou, K.; Su, N.; Petersen, H.; Stöcker, H.; Wang, X.-N., An equation-of-state-meter of quantum chromodynamics transition from deep learning. *Nature Communications* 2018, 9 (1), 1–6.
11. Braun-Munzinger, P.; Stachel, J., The quest for the quark–gluon plasma. *Nature* 2007, 448 (7151), 302–309.
12. Udrescu, S.-M.; Tegmark, M., AI Feynman: A physics-inspired method for symbolic regression. *Science Advances* 2020, 6 (16), eaay2631.
13. Seif, A.; Hafezi, M.; Jarzynski, C., Machine learning the thermodynamic arrow of time. *Nature Physics* 2020, 1–9.
14. Lin, H. W.; Tegmark, M.; Rolnick, D., Why does deep and cheap learning work so well? *Journal of Statistical Physics* 2017, 168 (6), 1223–1247.
15. Garg, S.; Lou, J.; Jain, A.; Nahmias, M., Dynamic precision analog computing for neural networks. *arXiv preprint arXiv:2102.06365* 2021, 1—12
16. Stepp, R., *Multiple-input Four-quadrant Multiplier*. 1992, © 2004-2022 FreePatentsOnline.com, patent 5115409.
17. Killoran, N.; Bromley, T. R.; Arrazola, J. M.; Schuld, M.; Quesada, N.; Lloyd, S., Continuous-variable quantum neural networks. *Physical Review Research* 2019, 1 (3), 033063.

18. Lin, X.; Rivenson, Y.; Yardimci, N. T.; Veli, M.; Luo, Y.; Jarrahi, M.; Ozcan, A., All-optical machine learning using diffractive deep neural networks. *Science* 2018, 361 (6406), 1004–1008.
19. Carleo, G.; Cirac, I.; Cranmer, K.; Daudet, L.; Schuld, M.; Tishby, N.; Vogt-Maranto, L.; Zdeborová, L., Machine learning and the physical sciences. *Reviews of Modern Physics* 2019, 91 (4), 045002.
20. Shen, Y.; Harris, N. C.; Skirlo, S.; Prabhu, M.; Baehr-Jones, T.; Hochberg, M.; Sun, X.; Zhao, S.; Larochelle, H.; Englund, D., Deep learning with coherent nanophotonic circuits. *Nature Photonics* 2017, 11 (7), 441.
21. Li, B.; Gu, P.; Shan, Y.; Wang, Y.; Chen, Y.; Yang, H., RRAM-based analog approximate computing. *IEEE Transactions on Computer-Aided Design of Integrated Circuits and Systems* 2015, 34 (12), 1905–1917.
22. Chuang, I. L.; Gershenfeld, N.; Kubinec, M., Experimental implementation of fast quantum searching. *Physical Review Letters* 1998, 80 (15), 3408.
23. Chen, S. Chinese scientists claim breakthrough in quantum computing race. *Bloomberg* [Online], 2020. https://www.bloomberg.com/.
24. Schuld, M.; Sinayskiy, I.; Petruccione, F., An introduction to quantum machine learning. *Contemporary Physics* 2015, 56 (2), 172–185.
25. Biamonte, J.; Wittek, P.; Pancotti, N.; Rebentrost, P.; Wiebe, N.; Lloyd, S., Quantum machine learning. *Nature* 2017, 549 (7671), 195–202.
26. Gastegger, M.; Behler, J.; Marquetand, P., Machine learning molecular dynamics for the simulation of infrared spectra. *Chemical Science* 2017, 8 (10), 6924–6935.
27. Wang, D.; Wei, S.; Yuan, A.; Tian, F.; Cao, K.; Zhao, Q.; Zhang, Y.; Zhou, C.; Song, X.; Xue, D., Machine learning magnetic parameters from spin configurations. *Advanced Science* 2020, 7 (16), 2000566.
28. Park, W.; Thekiniath, R.; Velagapudi, T., *Gravitational lensing with generative adversarial networks*. digitalcommons.imsa.edu, 2020.
29. Gebhard, T. D.; Kilbertus, N.; Harry, I.; Schölkopf, B., Convolutional neural networks: A magic bullet for gravitational-wave detection? *Physical Review D* 2019, 100 (6), 063015.
30. Feder, R. M.; Berger, P.; Stein, G., Nonlinear 3D cosmic web simulation with heavy-tailed generative adversarial networks. *Physical Review D* 2020, 102 (10), 103504.
31. Greene, B., *Until the end of time: Mind, matter, and our search for meaning in an evolving universe*. Knopf: 2020.

32. Baggott, J., *Quantum space: Loop quantum gravity and the search for the structure of space, time, and the universe*. Oxford University Press: 2018.
33. Greene, B., *The elegant universe: Superstrings, hidden dimensions, and the quest for the ultimate theory*. American Association of Physics Teachers: 2000.
34. He, Y.-H., Machine-learning the string landscape. *Physics Letters B* 2017, 774, 564–568.
35. Antoniadis, I. *In physics of extra dimensions, Journal of physics: Conference series*. IOP Publishing: 2006; p 015.

3

THE RULES OF THE GAME

VOLKER KNECHT AND KILIAN HIKARU SCHEUTWINKEL

Due to the growing significance of AI for humanity, the public intellectual and historian Yuval Noah Harari calls "algorithm" the most important word of our time.[1] Which algorithms make up the toolbox of ML, and to which problems in physics may they be applied? The present chapter will give an overview. Further algorithms and details are described in the Supplement.

ML algorithms are designed to build statistical models based on training via a learning procedure. The latter can follow one out of three different schemes: *supervised learning* optimizes a function mapping an input to an output based on example input–output pairs. On the other hand, *unsupervised learning* discovers patterns from unlabeled data. Finally, *reinforcement learning* deals with "intelligent" agents taking actions in an environment to maximize their cumulative reward, realizing a balance between exploration (of unfamiliar territory) and exploitation (of current knowledge).[2] These three types of approaches, as well as dedicated algorithms, are described in the following.

DOI: 10.1201/9781003245186-4

SUPERVISED LEARNING

A procedure of this type is trained to construct a suitable mapping between two sets associating every element of the first set (*input*) to exactly one element of the second set (*output* or *label*). This function represents the model derived from a set of data.

CLASSIFICATION VERSUS REGRESSION

The input could be an image of an astronomic light source to be categorized as star, galaxy, or quasar; these possible categories would then form the set of possible outputs. Thus, in this case, an image is mapped to a certain category. This mapping is taught in a training procedure where the correct label or output is provided for each input from a representative training set.

The input would commonly be a vector; this is essentially a column of numbers and can be seen as a point in a high-dimensional space. Each component then corresponds to the coordinate in a certain dimension. When the input is an image in grayscale, each component would encode the tone of a pixel as a real number. If the image is a megapixel one, the space of possible inputs would have one million (10^6) dimensions.

In the example, the set of outputs is discrete, consisting of three elements called labels. In this case, we talk about a *classification* task. This scheme is also utilized in particle physics to select events of interest or to identify particles from the data.

Alternatively, the set of outputs could be continuous, like real numbers. A possible task here could be to determine the temperature where a certain molecular material undergoes a transition from an amorphous solid to a viscous or rubbery state (denoted as *glass transition*), based on the knowledge from a set of materials studied previously. Here, the input could be selected continuous descriptors based on aspects of the electric charge distribution of the molecules, and the output would be glass transition temperature,[3] a positive real number and thus continuous. Such a continuous output characterizes a *regression* problem.

In such a type of problem, the output may even be multi-dimensional. Take the example of AlphaFold predicting the three-dimensional structure of a protein based on its amino acid sequence, using data of proteins whose structure is known. In this case, the output consists of the spatial coordinates of all non-hydrogen atoms involved, for a protein of an average size a vector with on the order of 10,000 dimensions.

SIMPLE MAPPINGS

For some methods, the mapping from new input data to output data can be determined directly from the training set. Procedures of this kind are termed *non-parametric methods*; an example is the *k*-nearest-neighbor algorithm described in the Supplement. Alternatively, the mapping can be expressed as a mathematical function depending on parameters chosen so as to reproduce the true outputs as closely as possible for the training set. Such techniques are called *parametric methods*. For the simpler of them, the parameters can be determined from the training set using a closed formula; this is, for example, the case for *linear regression* and *ridge regression*, two procedures described in the Supplement.

COMPLEX MAPPINGS

More complex mappings are usually implemented using artificial neural networks (see Figure 3.1) and often deep learning. A type of neural networks frequently used are feed-forward neural networks (see Figure 3.1A). Here, the information moves in only one direction – forward – from the input nodes.

The parameters of neural networks have the meaning of weights associated with the connections. Often, thousands or even millions of them are required, to be able to express complex relationships between the outputs and inputs. Suitable values for these parameters need to be determined numerically.

But what does "suitable" mean here? Well, a regression model should yield an output close to the true output, while a classification

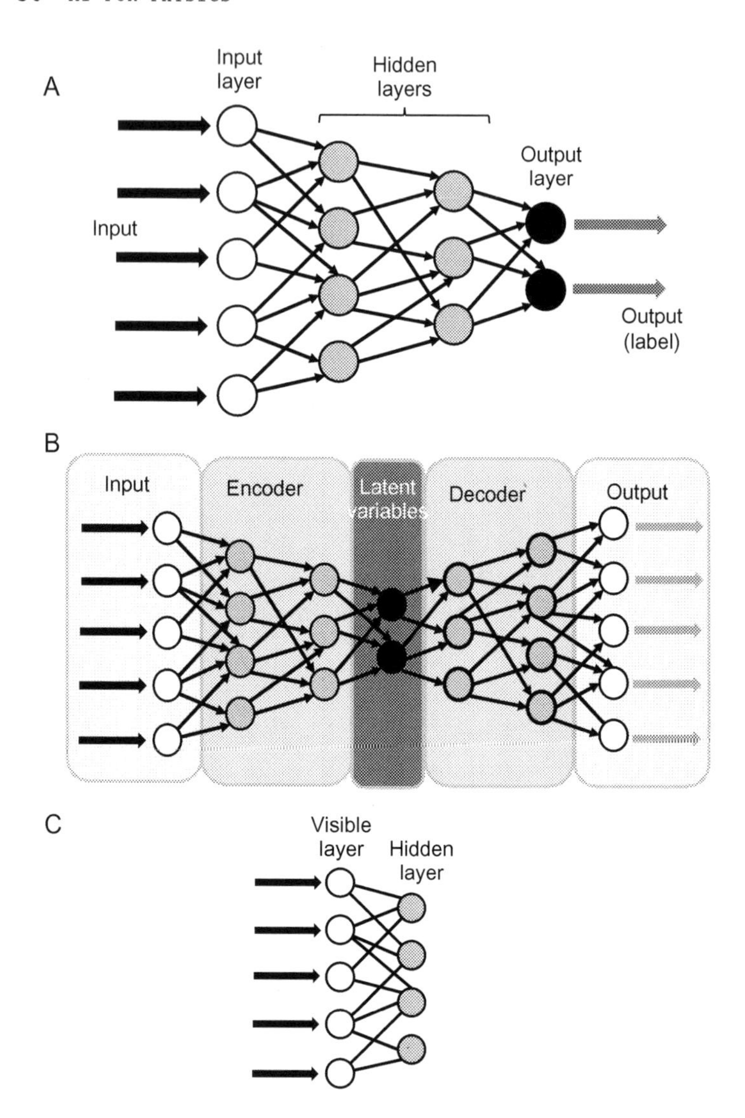

Figure 3.1 Artificial neural networks: (A) feed-forward neural network, (B) autoencoder, and (C) restricted Boltzmann machine.

model should generate a category matching the true one with high probability. Hence, a corresponding measure for the quality of an ML model is defined. This measure is denoted as *loss function*, and the value of this function for a particular input is called *loss*.[4]

In the case of a regression problem, a common loss function is the sum of the square deviations between the predicted and the true values of all components of an output vector. This loss depends on the model and the individual input. The loss averaged over a certain set of possible input data is denoted as *empirical risk*. The one for the training data is denoted as *training error*.[4] This refers to a regression problem. In the case of classification, on the other hand, the empirical risk may be defined as the fraction of times the learned function categorizes the data point incorrectly.[4]

To minimize the empirical risk (and thus optimize the model) then is the aim of the training procedure. Starting from a set of arbitrary weights, the latter are adapted in an iterative manner, until an optimal set of weights is obtained. But how does the algorithm know which weights to change in what way?

RIVER DEEP - MOUNTAIN HIGH

Well, to explain the idea we will first consider an amazingly simple functional mapping fully described with two weights only. Each pair of weights is then associated with an empirical risk. This is analogous to a geographical landscape with mountains: the weights resemble the location in two dimensions, in terms of the longitude and latitude, and the empirical risk is the corresponding height. Like Ike and Tina Turner's "River Deep – Mountain High", the risk landscape exhibits hills and valleys – mathematically spoken maxima and minima. Searching for optimal weights during the training procedure means starting somewhere in this environment – typically half-way up on a mountain (maximum) – and finding the way down to a valley (minimum).

How could you achieve this if you were blind? Now, you could fumble with your white cane to probe the rise or descent around you and take a step in the direction of steepest descent. Repeat this until the landscape around you is flat in all directions. This is the idea of gradient descent (GD).[5]

Now consider ML algorithms involve, typically, not only two but many more weights. These span a high-dimensional space, on which the empirical risk is defined. The corresponding optimization procedure is analogous to the low-dimensional example given above. Although it is hard to imagine, mathematics makes it possible.

This approach implies an iterative scheme where the descent in each direction is computed in each step, an often awfully expensive task. To reduce the computational cost at every iteration, only a random subset of the training set is used for evaluating the gradient. This technique is called *stochastic gradient descent* (SGD).[5] It can be applied to optimize neural networks with millions of parameters.

In principle. But various other phenomena occur in high-dimensional spaces not appearing in low-dimensional situations such as the three-dimensional physical space of everyday experience.

This issue is called *curse of dimensionality*. It means care must be taken not to construct a model containing more parameters than can be justified by the data, a problem denoted as *overfitting*.[6] This implies that an enormous amount of training data is needed to guarantee that there are several samples with each combination of values in the different dimensions. A rule of thumb is there should be at least five training examples for each dimension in the representation (i.e., for each parameter). Inversely stated, the number of parameters should be substantially smaller than the amount of input data.

CHOOSING THE NUMBER OF PARAMETERS AS A BALANCING ACT

If the number of parameters is too small, though, another problem occurs. Namely, a notorious problem of SGD and GD is they may not find the absolute minimum of the empirical risk (equal to zero) but get stuck in local or bad minima. This problem can be overcome by choosing a sufficiently large number of parameters, as this leads to conditions where any two points in parameter space with identical risk are connected by a path with constant risk.[7–9]

Interestingly, minimizing the risk in supervised learning is remarkably like minimizing the energy of a physical system, a well-studied problem, especially for glassy systems. Relating such a physical system to an ML model may yield deeper insights into the learning dynamics; the latter can show a sudden change with the number of parameters and may be mapped to a physical phase transition as described in the Supplement.[9]

It turns out that the number of parameters P needs to be large enough to avoid getting trapped in bad minima. At the same time, P must be small enough to avoid overfitting. Thus, P should achieve a subtle balance.

BIAS-VARIANCE TRADE-OFF

Once the parameters are optimized, the model derived must be tested on data not included in the training procedure to examine whether it is generalizable. To this aim, a second set of possible inputs, denoted as *test set*, is considered. The corresponding empirical risk is the expected error for an unseen sample, also called generalization error.[4]

This quantity can be decomposed into two different components. One of them is the *variance*. This observable is huge when the model is overly complex for the training set, thus suggesting trends in the data not existing, termed overfitting. The other error component is the *bias*; this quantity is large when the model is too simple. Hence, a good model is often a compromise between these opposite requirements; this *bias-variance trade-off* is a central issue in supervised learning.

KERNEL METHODS

Algorithms widely used for pattern analysis are kernel machines. The "kernel" involved in these methods refers to a similarity function over pairs of data points in raw representation. The different methods differ in the particular kernels used.

DECISION TREES

Algorithms which may be easier to interpret than kernel methods are *decision trees*.[10] They can be used both for classification and regression problems and are frequently applied in particle physics. A decision tree corresponds to a graph comprising nodes and connections (branches) in a tree-like manner. Mimicking a kind of quiz, each node is associated with a yes-or-no question to the input, the answer deciding whether to follow either of two branches originating from the node. The branch leads to another node, and so on, until reaching a leaf yielding the corresponding class or output value. Decision trees may be used to predict both continuous outcomes (regression trees) as well as discrete outcomes (classification trees).[4]

A complicated decision tree has low bias and high variance. The variance may be reduced if a whole ensemble of trees is considered, and the result is obtained via averaging over all the trees denoted as random forest.[11]

ARTIFICIAL NEURAL NETWORKS

Decision trees or random forests are restricted to relatively simple applications. Much more complex relationships can be mapped by *artificial neural networks* or simply *neural networks* (NNs) as highly non-linear systems. An NN is based on a collection of connected units or nodes called artificial neurons. Each connection can transmit a signal to another neuron. A neuron receiving a signal processes it and sends signals to neurons connected to it. The output of each neuron is computed by some non-linear mapping of the linear combination of its inputs plus an offset. Each of the corresponding coefficients is a weight associated with a respective connection and adjusting as learning proceeds.

The neurons are typically organized into multiple layers. Neurons of a given layer connect only to neurons of the immediately preceding and immediately following layers. The layer receiving external data is the input one. The layer yielding the result is the output one. In between them can be multiple hidden layers.

The latter are required to approximate a larger class of functions for the mapping of input data to labels than accessible otherwise. Even a single hidden layer with sufficiently many nodes enables the approximation of a very general class of functions (with low bias). The theorem, though, does not say whether this is possible with a feasible number of nodes.[12] Practical applications use many more hidden layers. For example, the AlphaFold neural net contains hundreds of layers with thousands of nodes.

Varying the number of hidden layers H, the bias of an NN is found to be maximal for a single hidden layer, while declining as H (and thus the complexity of the NN) is increased. On the other hand, the variance is minimal for a single hidden layer and first increases with H (as may be expected) to then (possibly unexpectedly) decrease again slowly. As a result, the expected generalization error is maximal for a single hidden layer and falls monotonously with increasing H.[13]

Multilayer perceptrons usually are fully connected networks whose major advantage is they do not require special assumptions about the input. While this renders fully connected networks very universally applicable, such networks do tend to have weaker performance than special-purpose networks tuned to the structure of a problem space. Making the explicit assumption that the inputs are images allows the encoding of certain properties into the model architecture. Here, at least one of the activation functions corresponds to the mathematical operation of a convolution, acting like a filter in image processing and implying most connections possible between the corresponding layers are removed; such a scheme is denoted as *convolutional neural network* (CNN).[14] This is a way of decreasing the number of parameters, reducing the problem of overfitting present for fully connected networks.

Neural networks may exhibit temporal dynamic behavior when the outputs of units feed back to the input in the next time step, characterizing a *recurrent neural network*.[15] The computation in such networks can be mapped to the dynamics of wave physics, as found very recently, paving the way for a new class of analog machine learning platforms.[16]

TREATING UNCERTAINTY AND PRIOR KNOWLEDGE: BAYESIAN INFERENCE

The methods above are based on choosing the optimal parameters of a model by minimizing a loss function. They correspond to the frequentist approach to statistical inference in which model parameters and hypotheses are fixed. Nothing is said about the statistical uncertainty of the parameter estimates. This is different in Bayesian inference where uncertainty in inferences is quantified using probability.[10] Parameters A are represented as random variables. Their probabilities $P(A|D)$ are updated after more evidence D is obtained or known using Bayes' theorem.

The quantity $P(A|D)$ is called posterior probability. According to Bayes' theorem it is proportional to two factors: one of them is the probability of the evidence D for a given parameter set A, $P(D|A)$; it is denoted as likelihood. The other one is the probability of the parameter based on prior knowledge and not considering the evidence, $P(A)$; it is called prior probability or short prior. The posterior probability $P(A|D)$ is furthermore normalized such that summing it over all possible parameter sets yields unity.

The most probable set of parameters is the one maximizing the posterior probability $P(A|D)$. Choosing it as a point estimate for the set of parameters A for a uniform prior $P(A)$ corresponds to a maximum likelihood estimate. It can be shown that the negative logarithm of the likelihood corresponds to the sum of the square deviations between the predicted and the true values of all components of an output vector, which is a typical loss function as noted above. This means that minimizing the loss function as in the frequentist approach corresponds to a maximum likelihood estimate or maximizing the posterior probability assuming a uniform prior.

But the Bayesian approach is more powerful, as it allows (A) previous knowledge to be incorporated via the prior and (B) the uncertainty of the set of parameters to be specified based on probability. The latter implies that instead of giving a point estimate for the parameter set a whole region containing the true parameters with a predefined probability may be specified.

For much of the 20th century, Bayesian methods were regarded as unfavorable by many statisticians for philosophical and practical reasons. Many Bayesian techniques were computationally expensive, thus most of the methods widely employed over the century were based on the frequentist interpretation. With the advent of powerful computers and new algorithms such as Markov Chain Monte Carlo, Bayesian methods are increasingly utilized in statistics in the 21st century. More details on Bayesian inference are given in the Supplement.

SYMBOLIC REGRESSION

The supervised learning tasks discussed so far consisted of classification and conventional regression techniques. The latter seek to optimize the parameters for a pre-specified model structure. A more general approach is *symbolic regression* (SR) inferring the model from the data. In other words, it attempts to discover both the structure and the parameters of the underlying model. SR searches the space of mathematical expressions to find the model best fitting a given data set, both in terms of accuracy and simplicity.

This technique was used by Udrescu and Tekmark to successfully re-discover 100 equations of *The Feynman Lectures on Physics*,[17] as described in Chapter 2.

UNSUPERVISED LEARNING

These learning schemes capture patterns in a self-organized manner.

CLUSTERING AND PRINCIPAL COMPONENT ANALYSIS

A classic example of unsupervised learning is data clustering where the data points are assigned to groups such that every group shares common properties. A common method is k-means clustering aiming to partition n input vectors into k clusters where each input belongs to the cluster with the nearest centre.[12] k-means clustering minimizes within-cluster variances (squared Euclidean distances).

Another type of data reduction is *principal component analysis*. It reduces the dimensionality of a problem by analyzing the correlations among the individual components of the input vectors. Hence, mutually independent collective descriptors are inferred as linear combinations of the original components. Thus, a few collective variables may be deduced, allowing a description of a system in a space with much lower dimension.

AUTOENCODERS

The outcome of a PCA is a reduced data set, from which the full data may be recreated. This restoration is associated with a reconstruction error appearing not to be optimal for PCA; a method associated with a seemingly smaller reconstruction error is an *autoencoder* (depicted in Figure 3.1B).[18] This object is an artificial neural network used to learn efficient data codings (in an unsupervised manner). The aim is to learn a compressed representation or encoding for a set of data, by training the network to ignore signal "noise". The simplest form of an autoencoder is a feed-forward non-recurrent neural network using an input layer and an output layer connected by one or more hidden layers. The output layer has the same number of nodes as the input layer, while the hidden layers have fewer nodes, providing a bottleneck for the information contained in the data. The goal is to reconstruct the input in the output to minimize the difference between them. As noted, the basic idea of an autoencoder is to reduce the dimension of the data (without losing essential information), enabling the improvement of performance on tasks such as classification.

PHYSICS-INSPIRED ALGORITHM: RESTRICTED BOLTZMANN MACHINE

Algorithms strongly inspired by physics are special neural networks called Boltzmann machines and especially restricted ones (RBMs, an example is shown in Figure 3.1C).[19] They are named after the Austrian physicist Ludwig Boltzmann (1844–1906) whose most important achievements were in the fields of thermodynamics and statistical mechanics. The Boltzmann statistics gives the probability of finding a given physical system in a certain state when it is in thermal equilibrium with a heat bath. This probability decreases exponentially with the energy of the state. RBMs can be used to infer the probability distribution of a high-dimensional observable given a training set of the observable. A detailed description of the algorithm is given in the Supplement.

GENERATIVE ADVERSARIAL NETWORKS

But is a single neural network always enough, or could multiple networks do a better job in developing a good statistical model from given data? If so, how could they work together to develop a statistical model? Should they cooperate or compete? What could this look like? Could it be a game? What could be the rules? The success of ML algorithms in chess was because two computers could play against each other very often in a short amount of time, thus gaining experience and developing innovative strategies very quickly just by trial and error.

Indeed, two artificial networks playing a game against each other can be useful for data analysis as well. This is shown by an unsupervised learning algorithm based on *generative adversarial networks* (GAN).[20] They consist of two artificial (typically feed-forward) neural networks performing a zero-sum game, where the loss of one is the gain of the other. Their common goal is to create new data with the same statistics as the training set. Candidates are produced by a *generative* network and evaluated by a *discriminative* one. The agents have opposing objectives: while the discriminator aims to distinguish synthesized data from the true data distribution, the generator's aim is to "fool" its antagonist by supplying novel candidates not recognized as such by the discriminator.

REINFORCEMENT LEARNING

Reinforcement learning is a set of ML methods where an (artificial) agent independently learns a strategy to maximize the rewards it obtains. The agent is not shown the best action in each situation but receives a reward or a penalty at certain points in time. Details on this are given in the Supplement.

WHAT'S NEXT?

The following chapters will take an application perspective. They will describe physical questions and problems addressed using ML and how, starting with applications at the subatomic scale in the next chapter.

REFERENCES

1. Harari, Y. N., *Homo Deus: A brief history of tomorrow*. Random House: 2016.
2. Bonaccorso, G., *Machine learning algorithms*. Packt Publishing Ltd: 2017.
3. Zhang, Y.; Xu, X., Machine learning glass transition temperature of polymers. *Heliyon* 2020, 6 (10), e05055.
4. Carleo, G.; Cirac, I.; Cranmer, K.; Daudet, L.; Schuld, M.; Tishby, N.; Vogt-Maranto, L.; Zdeborová, L., Machine learning and the physical sciences. *Reviews of Modern Physics* 2019, 91 (4), 045002.
5. Ruder, S., An overview of gradient descent optimization algorithms. *arXiv preprint arXiv*:1609.04747 2016.

6. Hawkins, D. M., The problem of overfitting. *Journal of chemical information and computer sciences* 2004, 44 (1), 1–12.
7. Freeman, C. D.; Bruna, J., Topology and geometry of deep rectified network optimization landscapes. *arXiv preprint arXiv:1611.01540* 2016.
8. Cooper, Y., The loss landscape of overparameterized neural networks. *arXiv preprint arXiv:1804.10200* 2018.
9. Spigler, S.; Geiger, M.; d'Ascoli, S.; Sagun, L.; Biroli, G.; Wyart, M., A jamming transition from under-to over-parametrization affects generalization in deep learning. *Journal of Physics A: Mathematical and Theoretical* 2019, 52 (47), 474001.
10. Box, G. E.; Tiao, G. C., *Bayesian inference in statistical analysis*. John Wiley & Sons: 2011; Vol. 40.
11. Biau, G.; Scornet, E., A random forest guided tour. *Test* 2016, 25 (2), 197–227.
12. Teknomo, K., K-means clustering tutorial. *Medicine* 2006, 100 (4), 3.
13. Neal, B., On the bias-variance tradeoff: textbooks need an update. *arXiv preprint arXiv:1912.08286* 2019.
14. Mehlig, B., Artificial neural networks. *arXiv preprint arXiv:1901.05639* 2019.
15. Le, Q. V., A tutorial on deep learning part 2: Autoencoders, convolutional neural networks and recurrent neural networks. *Google Brain* 2015, 1–20.
16. Hughes, T. W.; Williamson, I. A.; Minkov, M.; Fan, S., Wave physics as an analog recurrent neural network. *Science Advances* 2019, 5 (12), eaay6946.
17. Udrescu, S.-M.; Tegmark, M., AI Feynman: A physics-inspired method for symbolic regression. *Science Advances* 2020, 6 (16), eaay2631.
18. Kramer, M. A., Nonlinear principal component analysis using autoassociative neural networks. *AIChE Journal* 1991, 37 (2), 233–243.
19. Hinton, G. E., Boltzmann machine. *Scholarpedia* 2007, 2 (5), 1668.
20. Wang, K.; Gou, C.; Duan, Y.; Lin, Y.; Zheng, X.; Wang, F.-Y., Generative adversarial networks: introduction and outlook. *IEEE/CAA Journal of Automatica Sinica* 2017, 4 (4), 588–598.

PART II

MACHINE-LEARNING THE WORLD FROM SUBATOMIC TO COSMIC SCALES

4

AI FOR PARTICLE PHYSICS

MARIO CAMPANELLI AND VOLKER KNECHT

What is the world made of? Well, the most fundamental objects are *quantum fields*.[1] A field is a quantity that has a value (magnitude and possibly a direction) for each point in space and time, like an electromagnetic field. How one or more fields interact with matter is described by classical field theory.

Quantum mechanics tells us a field can be excited in discrete states called quanta and representing point-like particles. Such quantum fields underlie both the fundamental interactions and the particles of matter themselves. Coupling between the fields allows the conversion of a particle into other particles as well as the production and annihilation of matter-antimatter particle pairs.

Quantum fields are described via *quantum field theory* (QFT).[1] It combines classical field theory, quantum mechanics, and special relativity. General relativity, though, is excluded.

The particles of QFT can meet and interact, either by emitting or absorbing new particles, deflecting one another, or changing type. The bits exchanged to transmit the interactions are virtual particles arising from transient quantum fluctuations and existing only for a short time.

QFT is the theoretical basis of the Standard Model (SM) of particle physics,[2] which is the current established understanding of the

DOI: 10.1201/9781003245186-6

smallest building blocks of matter and their fundamental interactions. The SM is described in the following.

THE STANDARD MODEL

What are the elementary particles that make up our world according to the SM? As shown in Figure 4.1, the SM contains 12 types ("flavors") of fundamental *fermions*, plus their corresponding antiparticles, as well as four elementary *bosons* mediating the interactions (strong, weak, and electromagnetic forces, respectively) and the Higgs boson giving mass to all particles.

Bosons and fermions differ in their intrinsic angular momentum or *spin*. The latter only occurs in tiny quanta which is the unit in

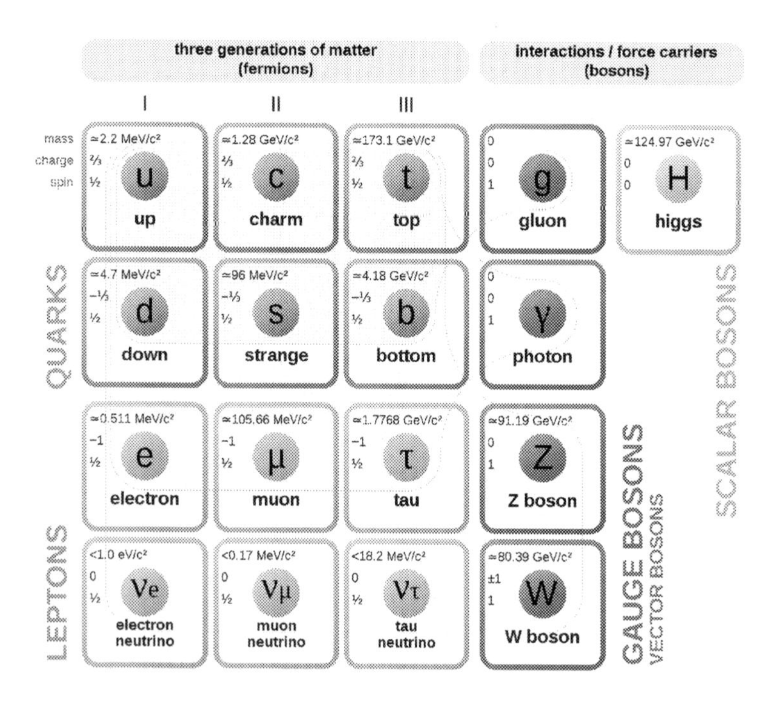

Figure 4.1 Standard model of elementary particles (by MissMJ, Cush; from Wikimedia Commons).

which particle spins are thus usually specified. The spins of bosons have integer and those of fermions half integer values. This implies fermions and bosons obey different statistics.

Bosons behave according to Bose–Einstein statistics, in which several indistinguishable particles can assume the same state. In contrast, fermions obey the Pauli exclusion principle *forbidding* identical fermions from occupying the same quantum state simultaneously. This rule is the ultimate reason why ordinary matter occupies volume.

The fermions are subdivided into *quarks* "feeling" the strong interaction and (much lighter) *leptons* which don't. Both the leptons and the quarks are split into three "generations" of one pair of particles each. The particles of each pair differ in their electric charge and their weak interaction. The leptons are the electron, the muon, and the tauon which are negatively charged, as well as their respective neutrinos which are neutral. The "flavors" of the quark generations are called up/down, charm/strange, and top/bottom.

At temperatures below two terakelvin, quarks clump together into elementary particles denoted as *hadrons*. Hadrons with an even number of quarks (mostly two – a quark and an antiquark) are called *mesons*. Hadrons with an odd number of quarks (mostly three) are denoted as *baryons*. Mesons and most baryons are very short-lived compared to *protons* and *neutrons* which are the most stable baryons and made up of the two lightest quarks – once up and twice down for neutrons but twice up and once down for protons.

Protons are positively charged while neutrons are electrically neutral. Both are called nucleons and make up atomic nuclei. Nucleons exchange π mesons which have a short lifetime, leading to a strong but short-range interaction between nucleons. Electrons are negatively charged and the most stable charged leptons; they form atomic shells.

The three fundamental interactions described by the SM play distinct roles in the existence of matter as we know it from everyday life. The strong interaction keeps atomic nuclei together while the weak force lets them decay. Finally, the electromagnetic interaction holds atoms and molecules, as well as condensed phases like solids and liquids together.

The SM has successfully explained almost all experimental results and precisely predicted a wide variety of phenomena. The last particle to be detected was the Higgs boson in 2012.

OPEN PROBLEMS

Nevertheless, lots of open questions remain. Some of them are described in the following. One concerns nature's ghost particles which are so difficult to detect: the neutrinos. Their mass is predicted as zero by the SM but appears nonzero in experiment.[3] Furthermore, the SM depends on 19 numerical parameters whose values are known from experiment but not explained by the theory.

An abyss of our ignorance opens when we look up to the sky. When we take stock, we should consider that the essence of the universe comes in two currencies whose exchange rate is found in Einstein's famous formula from special relativity: $E = mc^2$. It says that mass (m) and thus matter is equivalent to energy (E). The exchange rate is the speed of light (c). So how much of the energy of the universe is described by the SM? Well, apparently it is only a small fraction of five percent!

Why? Now, about 26 percent is *dark matter*.[4] This substance is not directly visible; however, its presence is suggested from its gravitational effect on the velocities by which visible stars orbit the center of their galaxy.

The remaining 69 percent of universe's energy is the so-called *dark energy*.[5] It corresponds to a vacuum energy and is indicated by the accelerated expansion of the universe against the effects of gravity.

The most profound problem of the SM is that it does not explain gravity, nor is it compatible with the most successful theory of gravity to date which is general relativity. The latter is based on classical physics. Missing is a *quantum* theory of gravity, which is the Holy Grail and biggest problem of modern theoretical physics.[6]

THEORIES BEYOND SM

Several theoretical approaches have been developed to overcome the problems of SM. Two of them are summarized in the following.

A speculated relationship between bosons and fermions is *supersymmetry*.[7] As a drawback, it involves many additional particles and parameters. But it among other things yields a dark matter candidate and provides a quantum theory of gravity.

In *string theory*, the point-like particles of the SM are superseded by one-dimensional objects called strings.[8] Bosonic string theory contained only bosons. In contrast, supersymmetric string theory – or *superstring theory* for short – accounts for both fermions and bosons and integrates supersymmetry to model gravity. This theory may also explain the observed forces and particles of the SM, justify their masses, and shed light on the nature of dark matter plus dark energy. In that sense, it provides a theory of everything (check out Chapter 8).

These approaches are highly intriguing but have not been verified to date.

MACHINE LEARNING PARTICLE PHYSICS

Experimental research in particle physics uses particle accelerators which propel particles to nearly the speed of light and bring them to collision. The largest collider in the world is the LHC where the Higgs boson was discovered, as mentioned in Chapter 2. The sensor arrays at the LHC produce data at a rate of about one petabyte per second.[9] A large portion of the measured signals are abandoned at the hard- and software level in real time before they are processed or permanently stored. The amount of data recorded is thus reduced to three gigabytes – still equivalent to several thousands of ebooks – per second.[10]

Given the huge size of data sets and that many analyses in particle physics start as classification problems, machine learning has been used in this field since the 1990s – mainly to improve the selection of relevant events to be stored for further analyses.

CUT-BASED EVENT SELECTION IN A PARTICLE PHYSICS EXPERIMENT

In a typical particle physics experiment, the signals from the detector are digitized, and corresponding numbers are processed by reconstruction codes and transformed into physical properties like energy and directions of the detected particles. Events from vastly different processes may produce very similar final states, and the first goal of an offline analysis is to separate the processes of interest – called signal – from the similar ones from a different origin – called background. Sophisticated Monte Carlo simulation codes are independently run to produce large samples of signal and background events to identify observables showing the largest separation power between signal and background events. The fraction of signal events in the sample is enhanced by only accepting events where the values of specific signal-enhancing observables are larger (or smaller) than a specifically chosen cutoff value, designed to retain most of the signal and to reject most of the background events. The problem of this approach is that the cutoff values for the various observables are independent of each other in that possible correlations between the observables are not considered, and events can be rejected even if only a single observable fails the cutoff values, even by a small amount. On the positive side, this technique is simple and fast, as well as easy to understand and implement.

PARTICLE AND EVENT SELECTION WITH NEURAL NETWORKS AND BOOSTED DECISION TREES

To improve over the cutoff-based event selection, one of the first applications of machine learning in particle physics was the use of feed-forward neural networks with only a few neurons. In this approach the input neurons are fed with observables that exhibit separation power between signal and background by having different distributions, and a single output neuron is used to discriminate the signal from a combination of backgrounds. More sophisticated networks exhibit multiple output neurons, in order to classify events into several categories. The ability of the networks to exploit hidden correlations between the variables and to combine them in a non-linear way allows improved selection performance with respect to cutting on individual observables, or even performing a simple linear combination.

Other commonly used ML tools for event classification are boosted decision trees. These techniques are an extension of the standard multi-dimensional cuts, with the difference that the events are not discarded if they fail just one cut but will go through other sets of cuts operating on the other variables, whose strength will depend on how many previous cuts have been passed, with the subsequent cuts being looser if the first ones have been passed or tighter otherwise. The sequence and values of the cutoffs are chosen automatically by an algorithm choosing a trade-off between efficiency and background rejection.

The performance of a binary classifier, like a neural network with a single output node, or a boosted decision tree aiming at separating signal from background events, is usually represented by a receiving operating characteristic (ROC)* curve, where, by varying the cut on the final classifier, the true positive rate is plotted versus the false positive rate. Other versions plot background rejection versus signal efficiency.

MACHINE LEARNING FOR JET PHYSICS

Hadronic jets are collimated sprays of particles produced by the formation of hadrons out of quarks and gluons denoted as hadronization. Due to the laws of strong interactions, particles with a color charge (the property leading to strong interactions) cannot live in isolation. At high energies, quarks will radiate gluons, which will

* A historical name coming from the development of radars during World War II.

produce quark–antiquark or gluon–antigluon pairs, in a process called parton shower. At lower energies, quarks will pick up other quark-antiquark pairs from the vacuum to produce more stable hadrons like pions, kaons, protons, and neutrons, which is what is called hadronization. When the initial quarks or gluons have sufficiently large momenta, the resulting hadrons will be collimated and all emitted in a narrow geometrical cone, the hadronic jet. Its axis, defined by the weighted average of the directions of the particles composing it, is a proxy for the direction of the quark or gluon that created it, and the jet transverse momentum a proxy for that of the initial particle. Defining which particles belong to a jet is not a trivial task, especially in situations where there are several nearby. Several jet clustering algorithms have been developed to, for instance, remove detector noise or calibrate the jets.

A very important area of jet physics that emerged in the last decade is unraveling the jet substructure:[11] this is done by developing techniques to look into the distribution of particles inside a jet and hence infer the particle producing the jet itself. A first classification is between jets originated by quarks and those originated by gluons. Although very similar, the fact that gluons interact with each other tends to lead to broader jets with more constituents. Even if a jet-by-jet separation of quark and gluon jets with large purity is not really possible because these jet classes are quite similar, some separation power on a large data set is still possible by combining the observables sensitive to this difference in a neural network or a multivariate classifier.

Hadronic jets can also come from the hadronic decay of heavy particles like the top quark, or the W, Z, or Higgs bosons. The hadronic decay of a top quark will produce three quarks generating the same number of jets, while a boson would decay into two quarks resulting in just as few jets. When the particles originating the jet have a very large momentum compared to their mass, they will receive a so-called Lorentz boost, and their decay products will all be reconstructed as a single jet. This jet, contrary to those originated by a single quark or gluon, will not have its highest radiation density close to its axis, but a multi-prong structure, with a cluster of high radiation density corresponding to the axes of the sub-jets relating to the quarks produced by the decay of the top quark or the bosons. Various observables have been constructed to distinguish jets with a single-prong structure coming from the simple hadronization of a quark or a gluon, from two- or three-prong, coming from hadronic decays of heavy objects. As these observables often exhibit correlations, combining them using a multivariate technique like principal component analysis provides better performance with respect to simple cutoffs.

The study of jet substructures has led to the use of much more sophisticated machine learning techniques than just the combination of observables using neural networks. The energy depositions of the particles in a jet can be treated like a picture, where for each bin the x and y positions correspond to the geometrical distance of the particle to the jet axis in the directions respectively normal and parallel to the beam, and the color corresponds to the transverse energy of each particle. Once a picture is made for each jet, a convolutional neural network – as a standard choice for image recognition – is used to distinguish between the different kinds of jets. Such networks will employ the information of the image at different scales, allowing the exploitation of various differences of jet production from their sources.

An even more sophisticated technique is the use of the Lund Jet Plane,[12] an abstract representation of the jet production mechanism on a two-dimensional plane. As mentioned in the beginning, the initial partons producing the jet will produce subsequent splittings and radiation emissions, the so-called parton shower. Even if it is not really possible to know the exact emissions at the parton level, it is possible to reconstruct it with some precision by clustering together particles in the jet being geometrically close and using that as an approximation for the emissions in the parton shower. Once its development has been reconstructed, a first possibility is to produce a Lund Jet Plane, where each splitting is represented by a point whose horizontal and vertical coordinates represent the momentum and angular separation of the splitting, respectively. Then a convolutional neural network can be used to separate the planes coming from signal and background jets. But the knowledge of the time ordering of the splitting contains more information than what can be included in the plane, and the performance in the identification of the particle originating the jet can be improved by including the history of the various splittings into a graph neural network.

CONVOLUTIONAL NEURAL NETWORKS FOR NEUTRINO EXPERIMENTS

Neutrinos are very light (but massive) particles, with a very small interaction with the rest of matter: being leptons they do not feel the strong interactions; being neutral they do not feel the electromagnetic interaction, and the gravitational interaction is negligible at these energies. So the only interaction acting on neutrinos in a sizable way is the weak one, leading to neutrinos being able to traverse large amount of materials with very small interaction probability. In a neutrino experiment, only a very small fraction of the neutrinos crossing the detector will interact with it, so in order to have a reasonable number of events, very large detectors are needed. The neutrinos themselves cannot be directly detected by any technology, but they are indirectly observed from their decay products.

Three kinds of neutrinos exist, called electron, muon, and tauon neutrinos according to the particle they produce when they have a charged current exchange with the detector material. Neutral current interactions, on the other hand, will produce undetectable neutrinos plus some hadronic activity in the final state, and they are identified from the absence of a charged lepton. One of the most used detector technologies for the detection of neutrinos in the energy range from a few MeV to several GeV is the Liquid Argon Time Projection Chamber (TPC).[13] The charged particles produced by neutrino interactions will ionize the argon, and the ionization electrons will be drifted in a strong electric field, leaving signals in planes of wires.

The combination of the signal position in the wires and the drift time allows the reconstruction of three-dimensional images of the events, and the amount of charge deposited by the tracks allows the identification of the detected particles. The complexity of the events, along with the large amount of data, makes the reconstruction of neutrino events in Liquid Argon TPCs the ideal testing ground for machine learning algorithms.

The MiniBooNE experiment at Fermilab (Chicago) used convolutional neural networks in many aspects of the analysis,[14] in a similar way as they are employed on jet images at the LHC. As neutrinos can interact anywhere in a large volume, the first step of the analysis is to define a bounding box, an area of the detector containing the particles produced in the interaction. This initial task is performed using a Faster-RCNN network,[15] which returns a

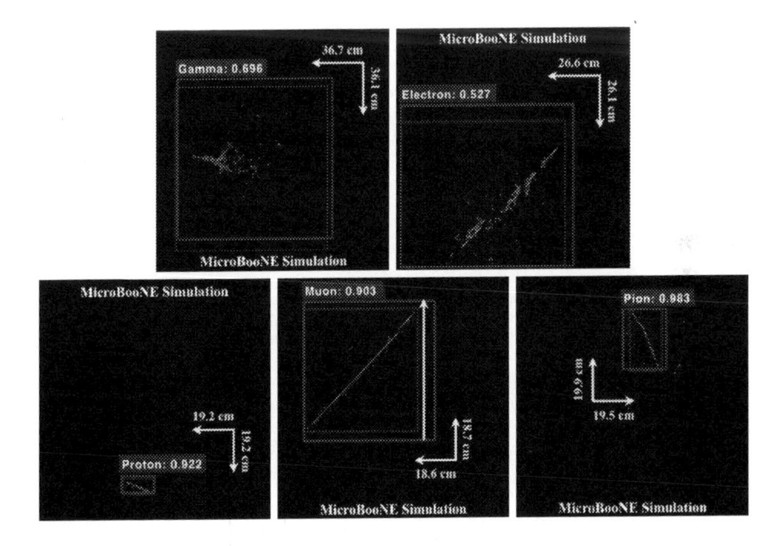

Figure 4.2 Machine learning neutrinos by detecting particles produced upon an interaction with the detector material: bounding boxes predicted by a convolutional neural network for five particle classes (© IOP Publishing Ltd and Sissa Medialab srl. Reproduced by permission of IOP Publishing.[14] All rights reserved).

number of boxes where the particles are expected to be contained. The images contained in these boxes, where the colors correspond to the energy deposition density of the particle, are passed to a multi-layer convolutional network, trained to distinguish five kinds of particles: electrons, protons, muons, pions, and photons. Typical images of these particles are shown in Figure 4.2. Electrons and photons interact in the detector producing electromagnetic showers (with different initial charge deposition), while the other particles are single tracks, with different energy deposition densities. The network is able to distinguish the various particle types with an accuracy of about 80 percent.

Neural networks are also used to separate neutrino events from a vast cosmic ray background (see Supplement), as another example highlighting the increasingly important role of machine learning the world at subatomic scales.

REFERENCES

1. Itzykson, C.; Zuber, J.-B., *Quantum field theory*. Courier Corporation: 2012.
2. Gaillard, M. K.; Grannis, P. D.; Sciulli, F. J., The standard model of particle physics. *Reviews of Modern Physics* 1999, 71 (2), S96.
3. Barger, V.; Marfatia, D.; Whisnant, K., *The physics of neutrinos*. Princeton University Press: 2012.
4. Einasto, J., Dark matter. *Brazilian Journal of Physics* 2013, 43 (5), 369–374.
5. Li, M.; Li, X.-D.; Wang, S.; Wang, Y., Dark energy. *Communications in Theoretical Physics*, 2011, 56, 525.
6. Baggott, J., *Quantum space: Loop quantum gravity and the search for the structure of space, time, and the universe*. Oxford University Press: 2018.
7. Martin, S. P., A supersymmetry primer. In *Perspectives on supersymmetry II*. World Scientific: 2010; pp 1–153.
8. Polchinski, J., String theory , Vol. 2, *Cambridge Monographs on Mathematical Physics*. **2005**.
9. Radovic, A.; Williams, M.; Rousseau, D.; Kagan, M.; Bonacorsi, D.; Himmel, A.; Aurisano, A.; Terao, K.; Wongjirad, T., Machine learning at the energy and intensity frontiers of particle physics. *Nature* 2018, 560 (7716), 41–48.

10. Hossenfelder, S., *Lost in math: How beauty leads physics astray*. Hachette UK: 2018.
11. Larkoski, A. J.; Moult, I.; Nachman, B., Jet substructure at the Large Hadron Collider: a review of recent advances in theory and machine learning. *Physics Reports* 2020, 841, 1–63.
12. Dreyer, F. A.; Salam, G. P.; Soyez, G., The Lund jet plane. *Journal of High Energy Physics* 2018, 2018 (12), 1–42.
13. Rubbia, C., The liquid-argon time projection chamber: A new concept for neutrino detectors. CERN: 1977.
14. Acciarri, R.; Adams, C.; An, R.; Asaadi, J.; Auger, M.; Bagby, L.; Baller, B.; Barr, G.; Bass, M.; Bay, F., Convolutional neural networks applied to neutrino events in a liquid argon time projection chamber. *Journal of Instrumentation* 2017, 12 (03), P03011.
15. Ren, S.; He, K.; Girshick, R.; Sun, J., Faster r-cnn: Towards real-time object detection with region proposal networks. *Advances in neural information processing systems* 2015, 28, 91–99.

5

AI FOR MOLECULAR PHYSICS

MAYANK AGRAWAL AND VOLKER KNECHT

Sharing the electrons of their outer shells – thus forming covalent bonds – atoms can combine to molecules of a range of sizes. This leads to a plethora of different substances. Over 350,000 chemicals and mixtures of chemicals have been registered for production and use. In addition, a wealth of synthetic polymers with different monomers of varying size, linear or branched, exist. There are also a huge number of biopolymers: carbohydrates, nucleic acids, and proteins.

The latter are chain molecules of various lengths, whose monomers are taken from an alphabet of 20 different amino acids as building blocks, whose sequence determines the three-dimensional structures of the molecules. Between 80,000 and 400,000 different proteins or polypeptides are estimated to be contained in the body of humans, one out of 1.6 million known and some estimated 8.7 million species on Earth, which implies a huge universe of different protein molecules.

Understanding the structure, dynamics, and function of proteins and other molecules, as well their mutual association, is the core of polymer, chemical, and biophysics, as well as molecular biology, chemistry, pharmacology, and materials science. The foundation of all these fields is molecular physics.

DOI: 10.1201/9781003245186-7

Experimental studies of molecules are complemented by theoretical investigations using models of varying levels of details. The most detailed treatment of a molecular system is based on quantum mechanics. Here the system's state is described by the wave function of all particles involved, yielding a value (complex number) for each combination of positions of the particles – a so-called *configuration*. The absolute square of the wave function is the probability density of measuring respective particle configurations. The relevant particles involved in a molecule are the corresponding electrons and atomic nuclei.

The time evolution of the wave function is described by the Schrödinger equation. It implies the change rate is proportional to the energy of the system.

To solve the Schrödinger equation exactly is only possible for very simple systems like a single hydrogen atom just involving one electron and one nucleus. To study systems of physical interest involving hundreds, thousands, or even millions of atoms, ab-initio calculations must use several approximations.

The first approximation considers that electrons are three orders of magnitude lighter than nuclei. Thus, it can be assumed the electron wave function immediately follows the nuclei positions. This is denoted as the *Born–Oppenheimer approximation*. Thereby, the Schrödinger equation splits into a still time-dependent equation for the wave function of the nuclei and a time-independent equation for the wave function of the electron for fixed nuclei positions. The latter yields the electronic ground state energy as a function of nuclei positions. This position-dependent energy inserts into the time-dependent equation for the wave function of the nuclei, yielding their interaction potential.

As a second approximation, the nuclei are treated as classical particles. Hence, the time-dependent Schrödinger equation for the wave function of the nuclei is replaced by the Newtonian equations of motion. Here, the forces on the nuclei are derived from the electronic ground state energy as a function of nuclei positions introduced above.

As a third approximation, the electronic degrees of freedom may not be described by the electronic wave function, but a simpler quantity hence derived – the electron *density* at all positions in space. The electron density determines the energy of the system; mathematically, the energy is a *function* of the electron density. The underlying concept is called *density functional theory* (DFT). This level of theory is the basis of *ab inito* MD simulations.

SPEEDING UP SIMULATIONS I: MACHINE LEARNING ATOMISTIC FORCE FIELDS

Ab-initio MD simulations are widely used to study molecules and material properties. Starting from an initial set of atomic coordinates and velocities, the Newtonian equations are solved in tiny (10^{-15} second) time steps. In each one, the interatomic forces are computed, and hence the accelerations, as well as the new velocities and positions of the atoms, are calculated. The time-consuming part is the computation of the forces. Calculating them using DFT is still highly expensive, such that systems comprising just a few hundred atoms can be simulated for timescales of the order of ten picoseconds only.

Over the past decades, many advances in the field allowed these calculations to be done progressively faster. First, the code has been gradually more parallelized such that many calculations are carried out simultaneously. Second, increasingly faster algorithms to solve the Newtonian equations of motion are implemented. Third, the DFT calculations may be accelerated using ML-based methods.

Here, the forces calculated via DFT are employed to train ML models which allow more efficient force calculations at later stages. The computational cost of ML model predictions is negligible compared to a DFT force-call, so replacing such a force-call by a reasonably accurate ML prediction reduces the computational expense significantly. Very recently[1] – and considering various examples – Yang and colleagues showed that using ML models can reduce the number of DFT force calls required to optimize the geometry of a system by 50 to 90 percent.

One of the most significant challenges to train such a model is that it requires the sampling of data across the whole configurational space of interest. To overcome this issue, an ML model is used to calculate properties for configurations close to configurations contained in a training data set while unseen configurations with high uncertainty (measured from the variation in the prediction from an ensemble of ML models or the deviation from regular DFT force calls) can be incorporated in the ML model by learning on-the-fly.[2] Due to the growing ML database, QM calls are needed less and less, and the calculations are increasingly ML-based and thus progressively more efficient.

A model successfully trained can reduce the computational expense dramatically, thus being able to cover time and/or length scales not practically feasible using *ab-initio* MD calculations. Recently, a team of researchers from the USA and China received the Gordon Bell Prize for developing a deep-learning algorithm that could simulate 100 million atoms for more than a nanosecond per day; this would take months or years if brute force DFT was used instead.[3] The increase in efficiency becomes even more apparent for high-throughput screening of a material database for a particular application. For example, Zhong and colleagues[4] predicted the energies for the adsorption of carbon dioxide (CO_2) molecules for more than 200,000 adsorption sites on more than 10,000 different surfaces employing a neural network model which reduced the computational expense by several orders of magnitude.

QM-parameterized ML models for MD simulations are often denoted as machine learning force fields. In general, a force field is a mathematical function yielding interatomic forces for a given configuration. Force fields have been employed for MD simulations for decades.

However, in the past, constructing accurate force fields was labor-intensive and required human effort and expertise. Machine learning force fields – which may be based on kernel methods or neural networks – may be parameterized in a fast and automatic way. Furthermore, ML-FFs are more flexible and typically more accurate than traditional force fields.[5]

USING MACHINE LEARNING TO ANALYZE OUTPUT OF SIMULATIONS

On the other hand, though, the latter have the advantage of being simpler in their mathematical form and allowing even more efficient calculations, being three orders of magnitude faster than even the fastest ML-FFs.[5] Conventional force fields are a sum of simple mathematical expressions involving two to four particles only. Here, covalent bonds are modeled as springs. Periodic angle potentials ensure only discrete bond torsion angles are allowed. Other potentials model interatomic repulsion due to the Pauli exclusion principle and electrostatic interactions of charged atoms. These parameters are fitted to data combining quantum-mechanical calculations and experimental measurements "by hand" in labor-intensive procedures.

Though these "hand-made" force fields involve fewer or no ML approaches for their development, their usage in MD simulations yields a wealth of data requiring ML methods for their analysis. On the other hand, these studies using these force fields also show the limitations of ML approaches; both will be discussed in the following.

Man-made force fields are extensively used to study biological molecules like proteins (shown in Figure 5.1). As noted above, the latter are chain-like molecules which fold into well-defined three-dimensional structures essential for their function as biological nanomachines. Their structure may be assessed experimentally via X-ray crystallography or nuclear magnetic resonance spectroscopy. A far more efficient but since very recently similarly reliable approach is to use machine learning; we recall the AI breakthrough in 2020 mentioned in Chapter 1: AlphaFold. While the latter may predict the final fold of a protein, though, it solves only half of the folding problem in that it shows the final structure but not the way to it. This demonstrates the limits of a purely data-driven approach such as AI.

Exploring folding dynamics requires adding physical insights as in the form of MD simulations. If the search for the current fold of a

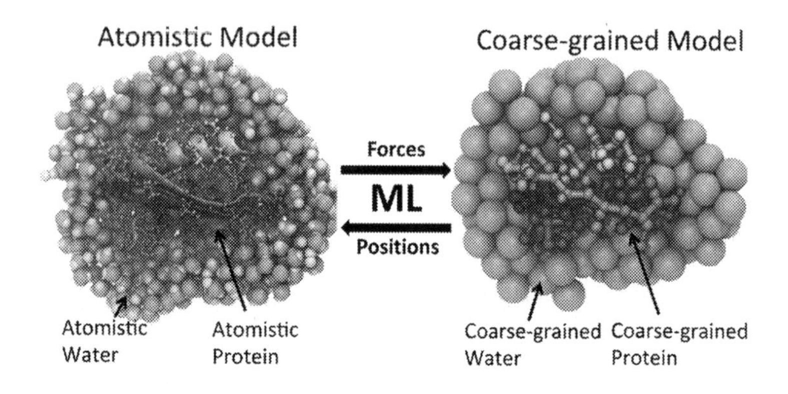

Figure 5.1 Machine learning molecular physics: inferring coarse-grained forces or atomistic configurations (image made with VMD by the Theoretical Computational Biophysics Group, University of Illinois at Urbana-Champaign, reproduced with permission).

protein was purely random, the time required for a protein of typical size to find its native state in the high-dimensional configurational space would exceed the age of the universe by ten orders of magnitude. Recent large-scale MD simulations on a special purpose supercomputer explain the short folding times observed experimentally by suggesting proteins tend to fold along dominant folding pathways.[6]

Analyzing conformational ensembles of proteins from MD simulations is not trivial. Proteins typically comprise thousands of atoms and (due to three-dimensional space) a three-fold number of degrees of freedom spanning a high-dimensional space of possible protein configurations.

Hence, analyzing data from MD simulations requires the dimensionality of the data to be reduced which can be done in various ways involving ML. Data from folding simulations are often investigated by clustering respective molecular configurations. Each cluster then corresponds to a protein *conformation*.

Accordingly, it is possible to map an ensemble of data points in a high-dimensional continuous space (configurations) to a discrete set of states (conformations). Each state is adopted with a certain probability, and transitions between a given state to any other one occur at a certain rate. Such ML-constructed so-called Markov models facilitate incredibly many short simulations of folding@home to be tied together.

Through distributed computing, this project employs the unused processing resources of personal computers and servers on which the software is installed, thus contributing to research on diseases. The ML-powered project achieved a combined computing power faster than the world's 500 fastest supercomputers combined on April 13, 2020, during the COVID-19 pandemic, outperforming the fastest supercomputer at that time by a factor of 15.[7]

Once a protein is folded it may fulfill its function. The latter depends not only on the protein's native structure but also its native *dynamics*, that is, the thermal fluctuations around its equilibrium state. To study these thermal fluctuations at unparalleled temporal and spatial resolution is possible via MD simulations. This may lead to an enormous amount of data whose analysis is very often centered around ML. Output data from MD simulations in terms of protein coordinates are taken as input of ML procedures, usually using unsupervised learning.

Here it is important to focus on *internal* motions. When cartesian coordinates (positions of each atom in each spatial dimension) are used this is achieved by translating and rotating each configuration back such that it fits on top of a reference configuration. As it is non-trivial to separate a rigid body from internal motions, it is usually more efficient to start from internal coordinates like distances or bond torsion angles. The resulting ensemble of configurations expressed in either way is a cloud of points in a high-dimensional space such that a clear interpretation of the data requires dimensionality reduction.

A method frequently employed to accomplish this is principal component analysis (PCA). Here, the covariance of the motion of each protein atom and the motion of any other protein atom in any direction (in the case of cartesian coordinates with removed rigid body motions) or (otherwise) any internal degree of freedom with any other internal degree of freedom is determined. This yields a huge quadratic array of numbers, the covariance matrix. Due to covalent bonds, excluded volume, and other interatomic interactions, the motions of two different atoms are often highly correlated. From the covariance matrix, though, a transformation may be constructed which reveals collective coordinates – called "principal components" – which are linear combinations of the original cartesian coordinates and mutually uncorrelated. Interestingly, usually on the order of only ten collective coordinates with the largest variances are found to show 80 percent of the overall variance. Focusing further analyses on these few softest modes allows an enormous dimensionality reduction.

PCA has the limitation to capture linear correlation only, while non-linear correlations may be detected via neural networks. Nevertheless, PCA can be superior to neural networks in some respects. Namely, comparing PCA with neural networks in terms of their ability to identify important amino acid residues in a protein based on their mobility, PCA tends to be both more accurate and computationally more efficient than autoencoders or restricted Boltzmann machines.[8]

PCA allows us to focus on collective coordinates with high variance as a spatial property. On the other hand, it is more efficient to concentrate on time-related characteristics when studying slow conformational transitions. This is accomplished by a modified variant of PCA which correlates coordinates not measured simultaneously but time-lagged relative to each other; this method is called time-lagged independent component analysis (TICA).

SPEEDING UP SIMULATIONS II: MACHINE LEARNING COARSE-GRAINED FORCE FIELDS

In the simulations described so far, every atom is described explicitly; therefore, these MD simulations are called atomistic. Despite all approximations involved they are still computationally expensive and thus limited in terms of the time and length scales accessible. A common approach is therefore to use a further simplification where multiple atoms are described by a single interaction site often called superatom. In addition, solvated (bio)molecules are often modeled without treating solvent explicitly (referred to as "solvent-free models"); rather, solvent effects on the dynamics of solutes are described implicitly using appropriate effective interaction potentials. Effects from the degrees of freedom of individual atoms not described explicitly may be treated implicitly using friction and random forces whose mutual balance determines the temperature of the system. This approach is the basis of coarse-grained (CG) MD simulations which are more efficient than atomistic simulations by two to three (or – as in the case of solvent-free models – even four) orders of magnitude![9, 10] CG MD simulations facilitate the study of systems of several hundreds of nanometers for time scales of micro- to milliseconds at a molecular level.[11]

The accuracy of CG force fields depends sensitively on the choice of parameters which thus need to be chosen carefully. Wang et al. proposed a ML scheme to parameterize a solvent-free CG model.[12] They used an artificial neural network mapping CG configurations to interparticle forces (as indicated in Figure 5.1); the model is constructed so as to reproduce the equilibrium distribution of the atomistic model, mapped to the CG coordinates. The training set is an ensemble of configurations from atomistic simulations. Furthermore, the loss function is defined as the square deviation of the total force on a CG particle between the atomistic and the CG model, averaged over all particles and all configurations. To describe the folding of a small polypeptide in water, the CG model was trained using atomistic simulations with an aggregate timescale of 200 microseconds. The folding behavior of the peptide was well reproduced by the CG model, as demonstrated from monitoring the joint dimensional distribution in a plane spanned by two collective variables derived from TICA. A lower bound of the prediction error arises from the noise level; the latter is due to discarding degrees of freedom in the CG model.

This CG model devised for a specific molecule is not transferrable to the study of other systems. To render the model transferrable, modifications like a dependence of the energy function on the CG particle types and their environment are necessary.[12]

Despite appearing relatively simple compared to biomolecular systems, pure water is still not fully understood and is the subject of intensive research. Chan and colleagues devised a CG model of water using machine learning; the model accurately describes the structure and anomalies of water and ice on mesoscopic scales while being two orders of magnitude cheaper in computational cost than existing atomistic models.[13]

On the other hand, solvent-free CG models combined with models with even lower resolution open the way to simulate whole biological cells at molecular detail. This was demonstrated recently by Pezeshkian and colleagues who combined dynamically triangulated surfaces for membranes with molecular descriptions.[11] By backmapping the surfaces to a molecular model, they were able to simulate the membrane of an entire mitochondrion – a cell organel enclosed by a double membrane and containing its own genetic material – in near-atomic detail. Even full atomic detail may be achieved via backmapping the CG model to an atomistic model. This task may efficiently be solved using ML regression models. In a recent study, Ana and Deshmukh demonstrated that using artificial neural networks, k-nearest neighbors, or random forests may yield better predictions than conventional backmapping approaches (see Figure 5.1).[14]

While the latter is an example of biophysics dealing with highly complex heterogeneous systems, the next chapter refers to more homogeneous systems whose investigation nevertheless has its own challenges which may be overcome by using machine learning methods: condensed matter physics.

REFERENCES

1. Yang, Y.; Jiménez-Negrón, O. A.; Kitchin, J. R., Machine-learning accelerated geometry optimization in molecular simulation. *The Journal of Chemical Physics* 2021, 154 (23), 234704.

2. Botu, V.; Ramprasad, R., Adaptive machine learning framework to accelerate ab initio molecular dynamics. *International Journal of Quantum Chemistry* 2015, *115* (16), 1074–1083.
3. Jia, W.; Wang, H.; Chen, M.; Lu, D.; Lin, L.; Car, R.; Weinan, E.; Zhang, L., Pushing the limit of molecular dynamics with ab initio accuracy to 100 million atoms with machine learning. In SC20: International Conference for High Performance Computing, Networking, Storage and Analysis, IEEE: Atlanta, GA, USA: 2020; pp 1–14.
4. Zhong, M.; Tran, K.; Min, Y.; Wang, C.; Wang, Z.; Dinh, C.-T.; De Luna, P.; Yu, Z.; Rasouli, A. S.; Brodersen, P., Accelerated discovery of CO 2 electrocatalysts using active machine learning. *Nature* 2020, *581* (7807), 178–183.
5. Noé, F.; Tkatchenko, A.; Müller, K.-R.; Clementi, C., Machine learning for molecular simulation. *Annual Review of Physical Chemistry* 2020, *71*, 361–390.
6. Lindorff-Larsen, K.; Piana, S.; Dror, R. O.; Shaw, D. E., How fast-folding proteins fold. *Science* 2011, *334* (6055), 517–520.
7. Köpf, A. Folding@Home übertrifft 2,4 Exaflops und ist damit schneller als die Top 500 Supercomputer. In *GameStar*, 14 April 2020 ed.; Webedia Gaming GmbH: Munic, 2020; https://www.gamestar.de/.
8. Fleetwood, O.; Kasimova, M. A.; Westerlund, A. M.; Delemotte, L., Molecular insights from conformational ensembles via machine learning. *Biophysical Journal* 2020, *118* (3), 765–780.
9. Monticelli, L.; Kandasamy, S. K.; Periole, X.; Larson, R. G.; Tieleman, D. P.; Marrink, S.-J., The MARTINI coarse-grained force field: Extension to proteins. *Journal of Chemical Theory and Computation* 2008, 4 (5), 819–834.
10. Arnarez, C.; Uusitalo, J. J.; Masman, M. F.; Ingólfsson, H. I.; De Jong, D. H.; Melo, M. N.; Periole, X.; De Vries, A. H.; Marrink, S. J., Dry Martini, a coarse-grained force field for lipid membrane simulations with implicit solvent. *Journal of Chemical Theory and Computation* 2015, *11* (1), 260–275.
11. Pezeshkian, W.; König, M.; Wassenaar, T. A.; Marrink, S. J., Backmapping triangulated surfaces to coarse-grained membrane models. *Nature Communications* 2020, *11* (1), 1–9.
12. Wang, J.; Olsson, S.; Wehmeyer, C.; Pérez, A.; Charron, N. E.; De Fabritiis, G.; Noé, F.; Clementi, C., Machine learning of coarse-grained molecular dynamics force fields. *ACS Central Science* 2019, 5 (5), 755–767.

13. Chan, H.; Cherukara, M. J.; Narayanan, B.; Loeffler, T. D.; Benmore, C.; Gray, S. K.; Sankaranarayanan, S. K., Machine learning coarse grained models for water. *Nature Communications* 2019, **10** (1), 1–14.
14. An, Y.; Deshmukh, S. A., Machine learning approach for accurate backmapping of coarse-grained models to all-atom models. *Chemical Communications* 2020, 56 (65), 9312–9315.

6

AI FOR CONDENSED MATTER PHYSICS

ÁLVARO DÍAZ FERNÁNDEZ, CHAO FANG, AND VOLKER KNECHT

The predecessor of condensed matter physics was solid-state physics. It attempts to understand the macroscopic characteristics of solids, including their thermal, electrical, magnetic, optical, and mechanical properties, starting from the microscopic constituents and their mutual interactions. One of the greatest achievements of solid-state physics was to clarify why some materials conduct electricity while others do not – distinguishing metals from insulators.

In 1968, Nobel laureate Phil W. Anderson advocated for a change in the name of solid-state physics to condensed matter physics.[1] The idea was to have a broader field encompassing non-solid matter as well, provided the interactions among its constituents were not negligible. The field encompasses exotic phases like the Bose–Einstein condensate composed of particles all occupying the same quantum state and thus showing macroscopic quantum behavior. An example of such a condensate is helium-4 at temperatures close to absolute zero where it forms a superfluid. Anderson also introduced a concept into physics which was borrowed from biology and touched upon briefly in the previous paragraph – emergence. It expresses the

DOI: 10.1201/9781003245186-8

idea that the whole is more than the sum of its parts, meaning each layer of complexity entails new properties which cannot be derived from the previous layer by brute force calculations involving the Schrödinger equation.[2]

USING MACHINE LEARNING TO OVERCOME SAMPLING PROBLEM FOR SPIN GLASSES

Our understanding of phases of matter was boosted by Lev Landau – a Nobel laureate in physics whose contributions permeate almost all areas of physics. According to Landau, phases could be classified according to their symmetry or, rather, symmetries spontaneously broken. The word spontaneous refers to the absence of external agents such as external fields. In going from a liquid to a solid, the continuous translational and rotational symmetries of the system are broken or reduced to discrete symmetries when the atoms in the solid occupy special positions to form a crystal. We say the solid is an ordered and the liquid a disordered phase.

A transition characterized by breaking rotational symmetry only occurs from magnetic interactions in solids as sketched in Figure 6.1A. Individual atoms constitute elementary magnets whose magnetic dipole moment mainly arises from electron spins. Neighboring elementary magnets ("spins") may favor a parallel alignment. The latter can give rise to a macroscopic alignment of spins – corresponding to a non-zero net magnetization – at temperatures below the so-called Curie temperature. This behavior called ferromagnetism is observed for metals like iron, cobalt, and nickel.

Many transition metals and their oxides favor an *anti*parallel alignment of adjacent spins. This phenomenon is termed antiferromagnetism and shown in Figure 6.1B.

Some materials show a mixed behavior comprising both ferro- and antiferromagnetic bonds which may also vary in strength due to positional disorder. Such a material is denoted as spin glass and sketched in Figure 6.1C.[6] It is not able to minimize the energies of all its atomic bonds simultaneously, giving rise to energetic frustration.

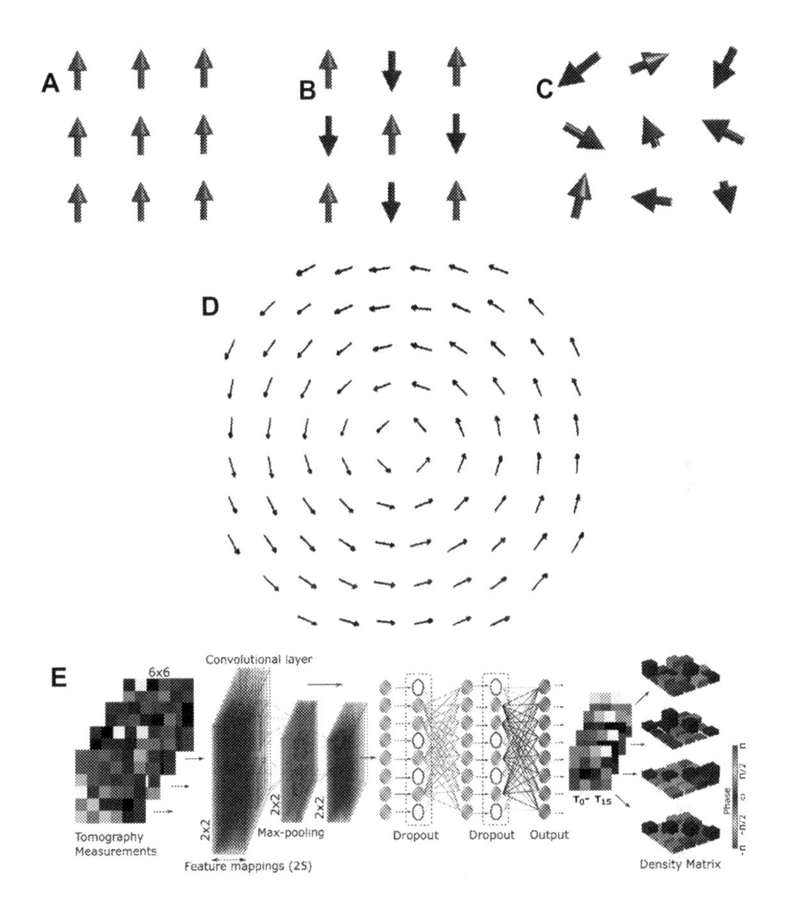

Figure 6.1 Machine learning condensed matter: (A) ferromagnet and (B) antiferromagnet, both periodically frozen (long range ordered). (C) Spin glass, randomly frozen (A–C are reproduced by permission of Rui Tamura[3]). (D) Vortex in Kosterlitz–Thouless transition (© IOP Publishing. Reproduced with permission.[4] All rights reserved). (E) Quantum tomography: using experimental measurements (left) to infer quantum states (density matrices of quantum systems, right) (reproduced from publication under CC BY license in *Machine Learning: Science and Technology*[5]).

Below a critical temperature all spins freeze cooperatively; the system gets trapped and may fluctuate between several states but cannot change to all other states of equivalent energy separated by high-energy barriers, whose height goes to infinity if the volume goes to infinity.[7] This behavior is like that of certain neural networks which is why studying spin glasses is also useful for understanding machine learning from a fundamental perspective.

Staying with physics, the temperature-dependent magnetic behavior of a material is often described using a model first investigated in more detail by Ernst Ising and denoted as the Ising model.[8] It describes a regular lattice of spins pointing up or down, with interactions between next neighbors governed by a uniform coupling parameter. A positive or negative parameter corresponds to a ferro- or antiferromagnetic behavior, respectively. Site-specific coupling parameters from a zero-centered bell-shaped distribution, on the other hand, yield the Edwards–Anderson model of a spin glass.

To yield a statistical ensemble of configurations of a model representative for a given temperature, spin configurations are sampled via MCMC simulations (see Chapter 3). Starting from a given configuration of the system, a new configuration is proposed and accepted with a probability which increases with temperature. The configurational change proposed may be flipping a single spin which is denoted as local update. This is fine for sufficiently high temperatures but approaching the phase transition temperature or even zero temperature leads to a notorious sampling problem which can render representative sampling of configurations unfeasible.

For ferromagnetic systems, this problem may be overcome by flipping not only single spins but whole clusters of aligned spins at once. Though such global-update methods also exist for spin glasses, novel and possibly more flexible MCMC methods would be extremely helpful. McNaughton and colleagues[9] proposed an unsupervised learning scheme in which smart global updates are performed using neural networks trained at higher temperatures to accelerate the sampling at lower temperatures. Using the two-dimensional Edwards–Anderson model (showing a spin glass transition at zero temperature) as an example, they showed that this scheme may accelerate the sampling compared to a local update algorithm by up to five orders of magnitude!

MACHINE LEARNING TOPOLOGICAL ORDER TRANSITION

A phase transition is well characterized from the behavior of a parameter characterizing the symmetry broken in one of the phases as a function of temperature; this parameter is termed order parameter. In the case of magnetic phase transitions, it is the net magnetization which is easy to define and evaluate. This is the case of phase transitions involving long-range order in one of the phases which does not apply to all phase transitions, as discussed in the following.

Mermin and Wagner showed in 1966 that a continuous symmetry could not spontaneously be broken in two dimensions or less and, therefore, there can be no long-range order at a finite temperature in that case. As a result, no phase transition can occur, and the system is expected to remain in a disordered phase.

Experimental evidence seemingly contradicting the Mermin–Wagner theorem indicated the possibility of a phase transition in two dimensions, most notably in thin films of helium-4. In view of this apparent inconsistency, in 1973 Kosterlitz and Thouless analyzed the problem using a generalization of the Ising model – the so-called XY or rotor model, where rotors living in a two-dimensional plane are free to rotate.[4] Each rotor represents the phase of the quantum-mechanical wave function at a given point in space. At zero temperature, the ground state corresponds to all the rotors aligned. Kosterlitz and Thouless showed that, at low temperatures, pairs of topological defects in the form of clockwise and counterclockwise swirls called vortices and antivortices, respectively, start to emerge; see Figure 6.1D. As the temperature is increased, the number of pairs grows, and so does their separation. The pairs do not disturb greatly the ordered ground state at zero temperature as they cancel out at large distances. As a result, the system is not long-range ordered, but *quasi*-long-range ordered.

As the temperature is increased further, there comes a point where the pairs unbind and isolated vortices and antivortices roam around the system which becomes a disordered plasma of vortices and antivortices. The change from bound to unbound pairs was later called Kosterlitz–Thouless transition. It is a topological phase transition because it is a transition from a disordered phase – although one with quasi-long-range order – to another disordered phase, driven by topological defects. Therefore, there is no violation of the Mermin–Wagner theorem and there is indeed a phase transition, although one not fitting into Landau's scheme. The Nobel Prize in Physics 2016 was awarded in part to Kosterlitz and Thouless for this discovery. The other part of the prize awarded to Thouless and Haldane concerns topological phases of matter, which we shall not touch upon here. The Kosterlitz–Thouless transition has been experimentally observed in several systems including the thin films of helium-4 and the melting of two-dimensional solids.

The classification of phases of matter and the detection of phase transitions pose a challenging goal and, among other techniques, machine learning has been put forward as a promising tool.[10] It has been implemented in scenarios comprising the Kosterlitz–Thouless transition. The task proves to be harder than the case of non-topological phase transitions. Neural networks have been implemented with the aim of learning vortices to detect the transition. It was suggested that correct classification of phases by convolutional networks requires feature engineering – selecting high-quality features with a high information content to avoid unnecessary data load.[11] The transition has also been correctly devised in a discrete version of the XY model, namely the clock model, using a fully connected neural network and training sets away from the transition temperature.[12] The input layer comprises correlation configurations[12] and improved correlation configurations.[13] Additionally, a study of the quantum XY model with training data extracted from the classical model has also been able to obtain the transition.[13] Finally, unsupervised algorithms have been applied to the XY model using non-linear dimensionality reduction methods called diffusion maps, successfully identifying the transition without the requirement of feature engineering.[14]

MACHINE LEARNING QUANTUM MANY-BODY SYSTEMS

Understanding systems made of many interacting particles – and requiring quantum mechanics to provide an accurate description of the system – is called the many-body problem. It is a problem of overwhelming complexity. The many-body problem encompasses a huge variety of systems generally not amenable to analytic solution and ranks among the most computationally intensive fields of science. One of the reasons for its difficulty, and arguably the most important one, stems from the vastness of the space of states, known as the Hilbert space, which increases exponentially with the number of particles involved. As a result, exact calculations are limited to small systems and require large amounts of computational power. Thus far, however, physicists have been able to explore many physical systems considering a single-particle approach and treating interactions in a perturbative fashion. Despite these successes, there are a great number of systems where quantum correlations and entanglement play a key role and, as a result, the single-particle approximation fails. Notable examples include high-temperature superconductors, whose origin is still unclear, but the most promising candidate theories rely on strong quantum correlations.

In order to address the many-body problem, several numerical methods have been devised, including *ab initio* methods such as density functional theory (see Chapter 5). Recently, a new player entered the stage, another approach to tackle the many-body problem: machine learning.[15, 16] One area where machine learning has proven to be particularly effective, even overcoming other standard methods, is obtaining neural-network representations of many-body wavefunctions, that is, the mathematical object containing all information about a many-body system. The result comprises so-called neural-network quantum states. Among other purposes, these states have been put forward to learn the ground state of a system, such as the Ising and Heisenberg models of magnetism.[17] Importantly, neural-network quantum states can be equipped with symmetries, the most challenging one being the exchange symmetry.[15] This is a symmetry that the many-body wavefunction has to fulfill upon the exchange of pairs of particles, giving rise to bosons and fermions (see Chapter 4). The main problem is that this kind of symmetry is not implemented in ordinary neural networks and, as a result, it is not possible to use the fully optimized networks now available for other tasks such as pattern recognition.

LOOKING FROM OUTSIDE: MACHINE LEARNING QUANTUM TOMOGRAPHY

A different view is taken by an experimentalist who performs measurements to reconstruct the quantum state of the system. This task is denoted as quantum tomography.[18] This approach, however, is limited to small systems with low entanglement. This limitation may be overcome via machine learning. Here – as illustrated in Figure 6.1E – experimental data are used to train a neural network via unsupervised learning to output the quantum state as a neural-network quantum state. Remarkably, the experimental samples can even be simulated numerically, meaning that by using conventional methods synthetic data can be obtained and fed into the network for training.

MACHINE LEARNING BASED DESIGN OF NEW MATERIALS AND QUANTUM STATES

Being highly relevant to technological applications, machine learning may be used to predict new materials with desired properties.[19] In particular, it is possible to predict thermodynamically stable compounds with a desired structure by searching over the entire compound space. Different approaches have proven to be successful at this task, examples being neural networks, kernel-based methods, and random forests. Similarly, new materials with a plethora of other properties, including critical temperatures, band gaps, and conductivities, can be obtained by means of machine learning which promises to become far more efficient than conventional methods.

Related to this design aspect is quantum state preparation which is part of the broader field of quantum control. The latter addresses the problem of engineering the quantum dynamics of a system by means of external fields, such as laser pulses or radio-frequency fields. Quantum state preparation in particular attempts to provide optimal or sub-optimal protocols to reach a desired quantum state from a given initial state under time evolution. Quantum control, originally conceived as a tool to control chemical reactions, is also currently being applied to a variety of areas, such as preparing Bose–Einstein condensates into target states or preparing quantum gates for superconducting qubits.[20] Finding a protocol which maximizes the fidelity – the overlap between the target and the time-evolved state – in a decent amount of time is a challenging task. Adiabatic (slow) evolution could be applied successfully for sufficiently long times; however, quantum systems are fragile and subject to noise and decoherence, meaning that a protocol should ideally be sufficiently fast to avoid these issues.[21]

Being an optimization problem, several gradient-based methods exist, such as gradient descent and stochastic gradient descent (see Chapter 3). Recently, however, more advanced machine learning methods have been applied to this problem. Specifically, most approaches rely on reinforcement learning algorithms, where the agent can learn the optimal protocol without any prior knowledge of the physical model.

Haug et al. considered a deep reinforcement learning algorithm with the fidelity as a reward. The goal is to be able to prepare arbitrary target states in a multilevel system by means of piecewise constant unitary evolution. The deep neural network is provided with an initial state and a random target state, and trained to yield the protocol parameters corresponding to the first unitary evolution.

Then, the evolved state and the target state are fed back into the network to obtain the second set of parameters. This is performed until an episode is completed with a fixed number of time steps. As a result, the network ends up learning clusters of protocols optimized in terms of protocol duration and driving strength. The algorithm is then applied to the case of the electron spin control in nitrogen-vacancy centers in diamond, a problem of interest due to its applicability in quantum sensing,[22] achieving high fidelities in as little as nine time steps or even fewer.

REFERENCES

1. Martin, J. D., *Solid state insurrection: How the science of substance made American physics matter*. University of Pittsburgh Press: 2018.
2. Altland, A.; Simons, B. D., *Condensed matter field theory*. Cambridge University Press: 2010.
3. Sato, S.; Uchida, Y.; Tamura, R., Spin symmetry breaking: Superparamagnetic and spin glass-like behavior observed in rod-like liquid crystalline organic compounds contacting nitroxide radical spins. *Symmetry* 2020, 12 (11), 1910.
4. Kosterlitz, J. M.; Thouless, D. J., Ordering, metastability and phase transitions in two-dimensional systems. *Journal of Physics C: Solid State Physics* 1973, 6 (7), 1181.
5. Lohani, S.; Kirby, B. T.; Brodsky, M.; Danaci, O.; Glasser, R. T., Machine learning assisted quantum state estimation. *Machine Learning: Science and Technology* 2020, 1 (3), 035007.
6. Binder, K.; Young, A. P., Spin glasses: Experimental facts, theoretical concepts, and open questions. *Reviews of Modern Physics* 1986, 58 (4), 801.
7. Parisi, G., Order parameter for spin-glasses. *Physical Review Letters* 1983, 50 (24), 1946.
8. Brush, S. G., History of the Lenz-Ising model. *Reviews of Modern Physics* 1967, 39 (4), 883.
9. McNaughton, B.; Milošević, M.; Perali, A.; Pilati, S., Boosting Monte Carlo simulations of spin glasses using autoregressive neural networks. *Physical Review E* 2020, 101 (5), 053312.

10. Carrasquilla, J.; Melko, R. G., Machine learning phases of matter. *Nature Physics* 2017, 13 (5), 431–434.
11. Beach, M. J.; Golubeva, A.; Melko, R. G., Machine learning vortices at the Kosterlitz-Thouless transition. *Physical Review B* 2018, 97 (4), 045207.
12. Shiina, K.; Mori, H.; Okabe, Y.; Lee, H. K., Machine-learning studies on spin models. *Scientific Reports* 2020, 10 (1), 1–6.
13. Tomita, Y.; Shiina, K.; Okabe, Y.; Lee, H. K., Machine-learning study using improved correlation configuration and application to quantum Monte Carlo simulation. *Physical Review E* 2020, 102 (2), 021302.
14. Rodriguez-Nieva, J. F.; Scheurer, M. S., Identifying topological order through unsupervised machine learning. *Nature Physics* 2019, 15 (8), 790–795.
15. Carleo, G.; Cirac, I.; Cranmer, K.; Daudet, L.; Schuld, M.; Tishby, N.; Vogt-Maranto, L.; Zdeborová, L., Machine learning and the physical sciences. *Reviews of Modern Physics* 2019, 91 (4), 045002.
16. Carrasquilla, J., Machine learning for quantum matter. *Advances in Physics: X* 2020, 5 (1), 1797528.
17. Carleo, G.; Troyer, M., Solving the quantum many-body problem with artificial neural networks. *Science* 2017, 355 (6325), 602–606.
18. Glasser, I.; Pancotti, N.; August, M.; Rodriguez, I. D.; Cirac, J. I., Neural-network quantum states, string-bond states, and chiral topological states. *Physical Review X* 2018, 8 (1), 011006.
19. Schmidt, J.; Marques, M. R.; Botti, S.; Marques, M. A., Recent advances and applications of machine learning in solid-state materials science. *NPJ Computational Materials* 2019, 5 (1), 1–36.
20. Glaser, S. J.; Boscain, U.; Calarco, T.; Koch, C. P.; Köckenberger, W.; Kosloff, R.; Kuprov, I.; Luy, B.; Schirmer, S.; Schulte-Herbrüggen, T., Training Schrödinger's cat: Quantum optimal control. *The European Physical Journal D* 2015, 69 (12), 1–24.
21. Bukov, M.; Day, A. G.; Sels, D.; Weinberg, P.; Polkovnikov, A.; Mehta, P., Reinforcement learning in different phases of quantum control. *Physical Review X* 2018, 8 (3), 031086.
22. Degen, C. L.; Reinhard, F.; Cappellaro, P., Quantum sensing. *Reviews of Modern Physics* 2017, 89 (3), 035002.

7

AI FOR COSMOLOGY

KILIAN HIKARU SCHEUTWINKEL, DANIEL GRÜN, BERNARD JONES, JIMENA GONZÁLEZ LOZANO, AND VOLKER KNECHT

How did our universe come into being, and how did it become the way it is now? The main ideas of the prevalent theory are summarized in the following.

THE CONCORDANCE MODEL OF COSMOLOGY

The predominant theory of the development of the cosmos is the Big Bang model displayed in Figure 7.1. Its fundamental feature is the creation of the universe from a state of superhigh temperature and density. The expansion of the universe and its emergence from a superhot dense state can be understood from the equations of general relativity.

The concordance model or standard model of Big Bang cosmology is the Lambda cold dark matter (ΛCDM) model where the universe contains three major components: dark energy, dark matter, and ordinary matter.[1]

DOI: 10.1201/9781003245186-9

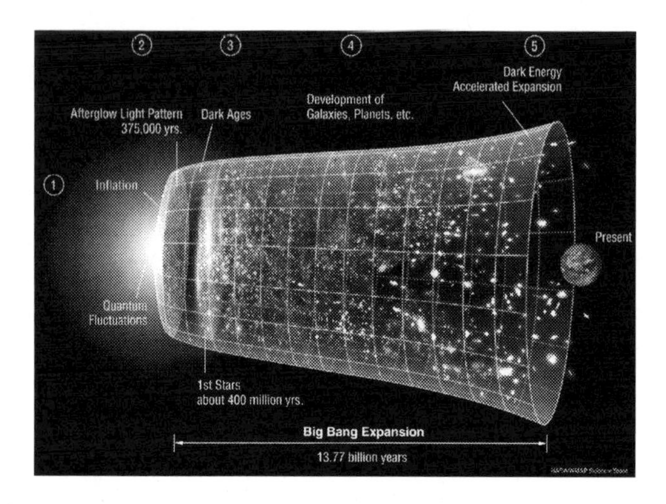

Figure 7.1 Standard model of cosmology (reproduced with permission from NASA/LAMBDA Archive/WMAP Science Team).

The expansion of space is obvious from the redshift of prominent spectral absorption or emission lines in the light from distant galaxies and the time dilation in the light decay of supernova luminosity curves. The redshift is attributed to the Doppler effect: a change in frequency of a wave in relation to an observer who is moving relative to the wave source.

The letter Λ represents the cosmological constant, currently associated with dark energy in empty space used to explain the contemporary accelerating expansion of space against the attractive effects of gravity (see Figure 7.1).

Dark matter is postulated to account for gravitational effects observed in very large-scale structures not accountable for by the quantity of observed matter. These effects include the speed at which visible stars orbit the center of their galaxy and the enhanced clustering of galaxies.

Cold dark matter (CDM) as currently inferred consists of particles beyond the Standard Model of particle physics. Furthermore, it cannot cool by radiating photons and interacts with other matter (be it ordinary or dark) only through gravity and possibly the weak force.

The "Big Bang" was the abrupt emergence of expanding spacetime containing radiation at temperatures of around 10^{15} K. This was followed by an exponential expansion of space by a scale multiplier of at least 10^{27} within only 10^{-29} seconds, a phase known as cosmic inflation (see Figures 7.1–7.2).

MACHINE LEARNING BIG DATA AND THE GLOBAL SHAPE OF THE UNIVERSE

The concordance model depends on six independent parameters fitted to experimental data; the latter include the age of the universe estimated as 13.8 billion years. Six further parameters are fixed; one of them is related to the total energy density (from dark energy as well as dark and ordinary matter) which determines whether the universe is spatially flat, open, or closed (corresponding to zero, negative, or positive curvature on a global scale, respectively). Bayesian analyses (see Figure 7.2A) suggest odds of the order of 50:1 or 200:1 in favor of a flat compared to a closed or open universe, respectively. The probability that the universe is spatially infinite depends on the underlying assumptions – i.e., the choice of the prior – and lies between 67 and 98 percent.[4]

Cosmological research of today and upcoming decades involves the world's largest telescope – the Square Kilometer Array.[5] It comprises three types of antenna elements; one of them is the SKA-Low array which alone will generate 5 zettabytes (10^{21} bytes) of data every year.[6] This is twice the data traffic in the world wide web in 2020 and – in terms of the number of bytes – five times the number of stars in the visible universe![7] Clearly, this huge amount of data needs to be processed automatically by a machine. Inversely, the sheer amount of these data is useful for machine learning algorithms which perform better the more (independent and identically distributed) data are available.

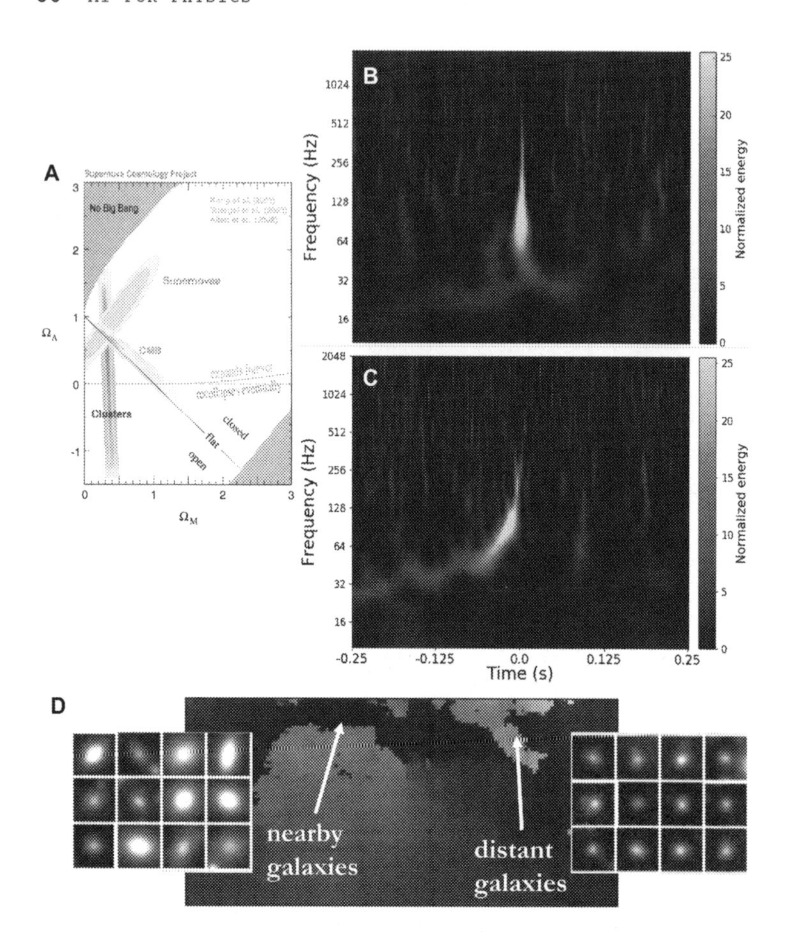

Figure 7.2 Machine learning cosmology. (A) Global shape of the universe: parameter space with relative densities of dark and ordinary matter (horizontal axis) versus the relative density of dark energy (vertical axis) with high probabilities from Bayesian analyses of three different types of experimental data (reproduced with permission from Jaan Einasto[2]). (B, C) Gravitational waves: distinguishing glitch (B) from signal (C) (B, C are reproduced with permission from the American Physical Society[3]). (D) Photometric redshift: a self-organizing map categorizes galaxies by color (here shown in grayscale). Because color is impacted by redshift, the resulting classes of galaxies also have similar distances from us.

MACHINE LEARNING NEW PHYSICS VERSUS INSTRUMENTAL EFFECTS

Machine learning may be useful in resolving a current debate triggered by recent radioastrophysical data from the lightest and by far most abundant element accounting for three-quarters of the mass of the ordinary matter: hydrogen. It formed 380,000 years after the Big Bang as protons and electrons recombined to form neutral atoms. Thus, there were no more free charged particles to elastically scatter photons of the thermal radiation from the 3,000 Kelvin hot "primordial gas", and the universe became "transparent". The thermal radiation from that time is still visible today, though due to the expansion of the universe in a redshifted manner and in the form of the cosmic microwave background radiation.

The recombination epoch was followed by the Dark Ages (see Figure 7.1) during which the clouds of hydrogen collapsed very slowly. This led to the formation of stars and galaxies which heralded the cosmic dawn about 300 million years after the Big Bang. The starlight contained photons with sufficiently high energies to kick electrons out from hydrogen atoms, thus reionizing them. This characterized the reionization epoch which persisted until a billion years after the Big Bang.

Clues to this era and the end of the Dark Ages are sought by observing neutral hydrogen's characteristic radio radiation. The latter arises from a flip of the spin of the electron relative to that of the proton and exhibits a wavelength of 21 cm. Today, the signal from these epochs is strongly redshifted to wavelengths of three to six meters. Measuring the 21-cm signal across the whole sky through radio antennas and comparing it with the cosmic microwave background allows us to infer how hydrogen matter evolved over time – to form the very first large structures of star-forming regions eventually grouping into the first galaxies of the universe. This hydrogen's radio radiation is known as the global signal and its strength depends on the quantity of neutral hydrogen present at a given time and the amount of background radiation in the universe.

Due to limitations in the sensitivity of radio antennas and computer power required to process the Big Data involved, these epochs are yet poorly explored. It is therefore impossible to discriminate between several competing astrophysical models, leaving high uncertainties in the minimal mass of star-forming haloes and various other parameters.[8] To date (as of November 2021), data for the global 21-cm signal during the cosmic dawn have been reported from a single experimental study only; the latter used a pair of table-sized spectrometer radio antennas in the Australian outback called the *Experiment to Detect the Global EoR Signature* (EDGES).[9] Remarkably, the signal described is more than twice as strong as predicted by current state-of-the-art astrophysical models[8] which raises an alert and a scientific debate[10, 11] around the analysis of the data from EDGES. Alternative models involve an enhanced radio background,[12, 13] instrumental effects yet unmodeled,[14, 15] or excessive cooling of the hydrogen clouds due to new physics including the interaction of hydrogen with dark matter.[14, 16] This debate may be resolved by including variations of these models in the nested sampling algorithm to infer which cosmological model is preferred by the data using Bayesian principles; this is ongoing research carried out by one of the authors of this chapter (K. S.).

Hydrogen's radio radiation is just one type of experimental signal allowing the probing of the universe. Other ones whose detection is facilitated via ML include gravitational waves (see Figure 7.2B,C) and the photometric redshift (see Figure 7.2D) as discussed below. Let us start with the latter.

MACHINE LEARNING PHOTOMETRIC REDSHIFT

OBJECTS IN THE MIRROR MAY BE BLUER THAN THEY APPEAR

Cosmology is the physical description of how our expanding universe changes its features over time. Due to the finite speed of light, as we look out to objects in the distance, we can observe the universe as it was in the past. For a measurement we make to contain information about cosmology, it has to relate some observable property of the universe, at a time in the past, to the size the universe had back then.

Fortunately, this size of the universe is imprinted into every single light quantum (or photon) inside of it. As the scale of the expanding universe increases, so do the wavelengths of photons which have traveled across it by the time they hit the mirror of our telescope. Because the stretching of wavelength causes an apparent shift of blue light toward the red part of the spectrum, this effect is called cosmological redshift. The 1:1 relation of the distance of remote galaxies to the redshift of the light we receive from them was the initial clue for Edwin Hubble's (1929)[17] and George Lemaître's (1927)[18] discovery that the universe is indeed expanding in all directions.

Redshift can be measured precisely if one is able to record the exact wavelength of each photon received (such as with a spectrograph, effectively a prism to turn each observed galaxy image into a rainbow). If one knows the original wavelength of the photons, as one does with spectral lines whose wavelengths are determined by atomic physics, the ratio of observed to original wavelengths directly determines the ratio of the size of the universe at the current time and the time the light was emitted.

However, even the most powerful telescopes in the world, with the most powerful spectrographic cameras attached to them, are so far unable to record spectra for as many and as faint galaxies as one would need to seriously challenge the current cosmological model. The Dark Energy Spectroscopic Instrument, leading the field since 2021, can achieve this for perhaps 30,000 galaxies on a typical night. The hundreds of millions or even billions of galaxies we need to observe for cosmological purposes can only be successfully recorded as images whose photons are counted in an integrated way across a handful of different wide "filters", i.e. ranges, of wavelengths. Their cosmological redshift influences the colors, i.e. the ratio of photon counts from filter to filter. But because no individual spectral lines can be determined, and because the broad shape of the spectra of galaxies shows great intrinsic variety, there is no unambiguous answer to what the redshift of every single galaxy is. This is the field and the problem of the so-called photometric redshifts.

AI TO THE RESCUE – BUT WITH THE RIGHT ARCHITECTURE AND TRAINING

Galaxies are extremely complex – no two galaxies are the same, and no model a human researcher could write down can fully describe the population of galaxies present in the universe. The relation between observed colors of galaxies and their redshift therefore lends itself to description by an artificial intelligence, trained on a subset of galaxies for which we have both observed their colors in sky images and determined their redshift with spectroscopy. Initial proposals for doing this were based on perceptron-style artificial neural networks trained to minimize the squared error of the estimated redshifts.[19] More recently it has been recognized that the intrinsic ambiguity of color for determining redshift means the estimated output of a machine for each galaxy needs to be an (often broad or multimodal) probability density of redshift, rather than a single number. The most important feature of this output is that when it is combined for millions of galaxies, the resulting distribution should be extremely close to the true distribution of their redshifts.

One thing this requires is that the training data are selected carefully and representatively. If, for instance, in the training data of spectroscopic redshifts one type of galaxy is overrepresented relative to another that shares the same color, or perhaps if one type of galaxy is altogether missing, that will make the output of the trained machine biased.[20] Similarly, if the training data contain galaxies whose redshifts have been determined incorrectly, the impact of that can render the results useless.

The need for ensemble redshift distributions in addition impacts the choice of machine learning architecture. Decision trees naturally produce probability distributions conditional on observed properties,[21, 22] and so do suitable variants of nearest-neighbor or kernel density estimates. A common issue is that while the space of observed colors and other properties is high-dimensional, galaxies only fill certain regions of that space. Techniques reducing the dimensionality of the observed data, e.g. unsupervised learning via self-organizing maps (see Figure 7.2D), are thus a common step for redshift estimation.[23]

More recently generative models are gaining traction as a suitable approach, including the first successful applications of autoencoder type architectures,[24, 25] but with potential also for generative adversarial approaches or restricted Boltzmann machines. Hybrid versions consisting of AI-based dimensionality reduction and classic Bayesian inference have already been put to work in the latest cosmological analyses.[26] If such models can be trained by the diverse existing data to represent both the true and the observed properties of the galaxy population, they could be the key to revealing the physical laws of the "dark" components of our universe from the billions of galaxies observed by the Vera C. Rubin Observatory and the Euclid space mission over the next decade.

MACHINE LEARNING COSMIC STRUCTURE

BUBBLE UNIVERSES ALL THE WAY DOWN

A fundamental assumption of cosmology is that our position and time are not special, but that the same laws of nature apply to the whole universe. One consequence of that is that every part of the universe evolves across time in many ways as a little bubble universe of its own, governed by the same laws of nature but influenced by statistical fluctuations of matter density that originate from quantum effects in the first tiny fraction of a second. Driven by the gravity of dark matter and the expansive effect of dark energy, the emptier patches expand rapidly into large voids, while the denser patches collapse and form structures like galaxies or galaxy clusters.

The formation and growth of structures over time is thus a most powerful diagnostic of the physics of dark matter and dark energy – essentially, it allows us to study their effect using many different bubble universes of varying matter density. However, how do we even measure these structures, given they are mostly made of invisible, dark matter?

DISTORTION PROBES GRAVITATION: INTERSTELLAR LENSING

The most immediate way to sense the presence of dark matter in the universe turns out to be subtle. General relativity predicts that the gravity of dark matter acts on the path of light rays the same way as on the motions of celestial bodies. Therefore, the distant galaxies we see are not where and what they seem: the positions, sizes, and most importantly the shapes of their images are distorted by the intervening matter distribution. You can work out the algebra of this to show that the effect is much like looking at the outside world through an uneven pane of glass – everything seems a little distorted, in a coherent way that depends on the variations in the thickness of the glass. This effect, called weak gravitational lensing, is our best bet for measuring the statistics of cosmic structure – but it is indeed weak. The images of galaxies appear elongated in a direction tangential around foreground masses by at most a few percent – and of course their shapes, positions, and distances on which the strength of this alignment depends are wildly varying intrinsically to begin with. The smallness of the effect, connected to how central it is to us understanding gravity in the cosmos, is a main motivator for the upcoming astronomical surveys that are going to take images of billions of galaxies.

Any signals of new physics that we can hope to find in these data would be very small, but none the less could change the whole of fundamental physics. Artificial intelligence may be the key for this next revolution in physics.

FISHING FOR COMPLEMENTS WITH THE COSMIC WEB

Once we have these painstakingly collected data and calibrated galaxy measurements, what should we then best use them for? This question is surprisingly difficult to answer as the distribution of matter density in the grown-up universe is highly complex – it is not just Gaussian (bell-curve shaped) fluctuations, and it is not just an ensemble of dark matter haloes. Rather, no single description can capture the "cosmic web" of filaments, haloes, and empty spaces fully. Each way of looking at them will offer complementary information, allowing us to test our model of physics more stringently, and in different ways. Cosmologists have historically relied on mostly just a single simple feature of this matter density field – the root-mean-square variation from one place to the other – to check their models. But that may make them miss the essential clues: general relativity, dark matter, and dark energy should allow us to predict the full picture of cosmic structure, not just one of its simple statistics.

Artificial intelligence offers a promising alternative: let the machine decide whether what we see in the universe matches our expectation or not. Several studies have developed the techniques necessary for that, and some are even already applying them to the data from weak lensing astronomical surveys. The geometry of these data, that essentially consist of maps of the sky, has led to the implementation of innovative AI concepts. For deep convolutional neural networks, the geometry of the pixelated sphere lends itself to a graph representation. Its high dimensionality, commonly with millions of independent pixels, makes the problem attractive for data compression methods, among them so-called deep summaries or information-maximizing neural networks. Generative models can be used to learn the diversity of cosmic structures and represent it in cosmological analyses. It is quite possible that one such architecture would give the first strong hints of a discovery of new physics from the cosmic web, given their superior ability to extract its information. That does not mean we would have to blindly trust the answer from the machine. Instead, we can learn from the artificial intelligence to form an optical statistical description of the data collected on cosmic structures, and trace what might lead a machine to see where modern physics so far has had a blind spot.[1]

MACHINE LEARNING GRAVITATIONAL WAVES

Gravitational lensing mentioned above arises from the static effect of a mass distribution on the curvature of spacetime. A *dynamic* effect from moving masses occurs if the masses' motion is not perfectly spherically or rotationally symmetric; perturbations in the curvature of spacetime appear which propagate in the form of gravitational waves (GW). They squash and elongate whatever gets in their way as they go by. Gravitational waves can pass through any intervening matter without being scattered significantly; they may therefore carry information about astronomical phenomena heretofore never observed by humans. These waves were predicted from general relativity in 1915.

One challenge in the experimental verification of this prediction was to detect a corresponding signal against an overwhelming noise background, similar to the detection of the God particle discussed in Chapter 2. In fact, recognizing a meaningful signal in noisy data is a very old problem going back to the time when observations of the positions of planets and comets in the sky were first recorded with enough accuracy to determine their orbits through the Solar System. The earliest example of this was surely Apollonius of Perga's proposal of an epicyclic theory for planetary motions, around 200 BCE, and its exploitation in the 2nd century AD by Ptolemy of Thiebaud in Egypt in his Almagest. Planetary positions were determined only approximately, yet he managed to overcome the inherent errors in the measurements and to make viable predictions for the future positions of the planets. That model, with some modifications, survived for 1,300 years, until the work of Johannes Kepler, working within the solar-centric framework for the planetary motions of Nicolas Copernicus.

What both the epicyclic theory and Kepler's elliptical orbits had done was to fit the planetary motions by a combination of two circular motions with different centers. There was no need to add more circles having higher circular frequencies: to do so would have been inconsistent with the accuracy of the measurements. In effect, they had used a technique called "Fourier analysis" to eliminate the "noise" from measurement errors via decomposing temporal or spatial signals to their frequency components.

The use of Fourier analysis in cleaning signals of noise came to be of central importance at the start of World War II with the appearance of radar. The primitive equipment of the time returned weak signals from reflections off approaching aircraft. The signals were buried in noise. At that point the mathematical process referred to as "fast Fourier transform" (FFT) was invented in secret and independently by a number of separate research groups working on radar. The first digital computers were coming online at this time. They had limited memory, they were by modern standards very slow, and they were difficult to program. But they were fast enough to handle the small FFT necessary to clean the radar signals.

This was in many ways the birth of data processing and, in particular, image and audio processing. Now, many of us have experienced the benefits of such technologies in noise-canceling earphones. We listen to speech and music in such a way that the outside world does not impinge on our senses. Our mobile phones also prettify our poorly exposed and noisy pictures – gone are the days of blurry and poorly exposed pictures not focused on the faces we are photographing.

Fundamentally, sound and light are wave phenomena. What we hear and see is a superposition of waves our brain has learnt to interpret. Sound is a superposition of a time-sequence of sound waves while images are a spatial organization of light waves of different frequencies (colors). The nature of the sound we hear is determined by the structure of the received packet of waves. A violin sounds different from a flute or the pronunciation of a vowel sound because the frequency make-up of the packets received is different in each case. The character of the sound is determined by the wave shape, referred to as a "formant". The task of speech recognition systems is to recognize these formants. Noise added to the sound distorts the formant shape, and so the more noise there is the harder it is to recognize the sound.

In this regard, sound waves are like gravity waves. The latter emitted by the merging of two black holes have a "signature" that depends on the properties of the merging objects, their mass and spin, and on their mutual orbit.

A century after their prediction, GWs were discovered experimentally in 2015 – by the LIGO gravitational wave detectors as a signal from the merger of two black holes. LIGO stands for Laser Interferometer Gravitational-Wave Observatory, and it consists of two observatories located in the US in Louisiana and Washington. Each setup starts with a powerful laser beam split into two perpendicular 4-km arms. Each light beam is reflected by a mirror located at the end of the arm, and they are combined again at the beam splitter. Finally, the combined signal is measured as an interference pattern by a light detector. The length of the arms is tuned such that at the detector the beam waves are out of phase; this means that the peaks of one light beam wave are added to the troughs of the other. This is called destructive interference; the waves cancel each other, and the detector does not measure any signal. When a gravitational wave passes through the interferometer the arms can be stretched or shortened, forcing the beams to be less out of phase, and a signal is detected.

The most energetic GW are produced by cataclysmic events such as explosions of massive stars or the merging of black holes. Even though these events release immense amounts of energy, their effects of contracting or expanding space are too faint to be easily observed. In 1969, scientists started contemplating possible designs for a device that could detect GW. However, it wasn't until 2015 that the LIGO collaboration was able to make the first direct GW detection.[27]

One important challenge LIGO faces is that GW detections can be faked by noise artifacts originated by the physical environment (seismic activity, complexities in the detectors) (Figure 7.2B,C). Identifying these glitches is crucial to reduce the number of false detections, and understanding their origin could lead to improving the detectors.[28] Currently, glitches are identified by an algorithm called Omicron, which identifies them by searching for power excess in the GW data stream.[29] Recently, researchers from the University of Columbia developed a new glitch-identification algorithm that only needs data from the "auxiliary channels".[3] These channels characterize the detector's components and the physical environment. Thus, this method can independently confirm if a signal is a GW signal or a glitch. The algorithm is based on logistic regression, a well-established machine learning method similar to linear regression. The difference is that a linear combination of the input components enters the logistic function (i.e., a smoothed step function). This forces the output to be a real number between 0 and 1 that can be interpreted as a probability in a two-class classification problem.[30]

This algorithm is the first glitch-identification method considering all the auxiliary channels, a computationally expensive task.[3] In a pre-processing step, all the channels not informative for glitches or that may be coupled with the GW data are removed. Previous Omicron results were used to create the training and testing samples for the two classes: glitchy times and glitch-free times, which were selected as windows of three seconds from the peak time of glitches and stretches without glitches, respectively. For each channel event point, ten features were hand-chosen based on intuition to describe the channel's behavior in both classes. For one of the analyses, the training data set was composed of 2,500 glitchy times and 2,500 glitch-free times, both samples drawn randomly from a 10,000-second period. During the model optimization, a term proportional to a constant was added to the solution to increase the bias and the number of zero coefficients, so the model is easier to interpret. One of the trained models has only 0.02 percent nonzero coefficients, which suggests only a small subset of channel features is necessary for glitch identification. The testing data consisted of

7,500 glitchy times and 7,500 glitch-free times, on which the model reached a classification accuracy of 79.9 percent when picking the glitch probability threshold at 0.5. This prediction threshold can be chosen to adjust the number of true positives and false negatives. In conclusion, this method can correctly identify the presence or absence of glitches, and it has the potential to be used in near-real-time analysis.

NOTE

1. Issues with calibrating ML methods are discussed in the Supplement.

REFERENCES

1. López-Corredoira, M., Tests and problems of the standard model in Cosmology. *Foundations of Physics* 2017, 47 (6), 711–768.
2. Einasto, J., Dark matter. *Brazilian Journal of Physics* 2013, 43 (5), 369–374.
3. Colgan, R. E.; Corley, K. R.; Lau, Y.; Bartos, I.; Wright, J. N.; Márka, Z.; Márka, S., Efficient gravitational-wave glitch identification from environmental data through machine learning. *Physical Review D* 2020, 101 (10), 102003.
4. Vardanyan, M.; Trotta, R.; Silk, J., How flat can you get? A model comparison perspective on the curvature of the Universe. *Monthly Notices of the Royal Astronomical Society* 2009, 397 (1), 431–444.
5. Dewdney, P. E.; Hall, P. J.; Schilizzi, R. T.; Lazio, T. J. L., The square kilometre array. *Proceedings of the IEEE* 2009, 97 (8), 1482–1496.
6. Pool, R. Drowning in data. https://spie.org/news/photonics-focus/may-jun-2020/square-kilometer-array-big-data?SSO=1 (accessed December 1, 2021).
7. Vuleta, B. How much data is created every day? [27 Staggering Stats]. https://seedscientific.com/how-much-data-is-created-every-day/ (accessed December 1, 2021).
8. Cohen, A.; Fialkov, A.; Barkana, R.; Lotem, M., Charting the parameter space of the global 21-cm signal. *Monthly Notices of the Royal Astronomical Society* 2017, 472 (2), 1915–1931.

9. Bowman, J. D.; Rogers, A. E.; Monsalve, R. A.; Mozdzen, T. J.; Mahesh, N., An absorption profile centred at 78 megahertz in the sky-averaged spectrum. *Nature* 2018, 555 (7694), 67–70.
10. Hills, R.; Kulkarni, G.; Meerburg, P. D.; Puchwein, E., Concerns about modelling of the EDGES data. *Nature* 2018, 564 (7736), E32–E34.
11. Bowman, J. D.; Rogers, A. E.; Monsalve, R. A.; Mozdzen, T. J.; Mahesh, N., Reply to Hills et al. *Nature* 2018, 564 (7736), E35.
12. Fialkov, A.; Barkana, R., Signature of excess radio background in the 21-cm global signal and power spectrum. *Monthly Notices of the Royal Astronomical Society* 2019, 486 (2), 1763–1773.
13. Jana, R.; Nath, B. B.; Biermann, P. L., Radio background and IGM heating due to Pop III supernova explosions. *Monthly Notices of the Royal Astronomical Society* 2019, 483 (4), 5329–5333.
14. Bradley, R. F.; Tauscher, K.; Rapetti, D.; Burns, J. O., A ground plane artifact that induces an absorption profile in averaged spectra from global 21 cm measurements, with possible application to EDGES. *The Astrophysical Journal* 2019, 874 (2), 153.
15. Singh, S.; Subrahmanyan, R., The redshifted 21 cm signal in the EDGES low-band spectrum. *The Astrophysical Journal* 2019, 880 (1), 26.
16. Barkana, R.; Outmezguine, N. J.; Redigolo, D.; Volansky, T., Strong constraints on light dark matter interpretation of the EDGES signal. *Physical Review D* 2018, 98 (10), 103005.
17. Hubble, E., A relation between distance and radial velocity among extra-galactic nebulae. *Proceedings of the National Academy of Science* 1929, 15 (3), 168–173.
18. Lemaître, G., *Un Univers homogène de masse constante et de rayon croissant rendant compte de la vitesse radiale des nébuleuses extra-galactiques*, Annales de la Société scientifique de Bruxelles, 1927; pp 49–59.
19. Collister, A. A.; Lahav, O., ANNz: Estimating photometric redshifts using artificial neural networks. *Publications of the Astronomical Society of the Pacific* 2004, 116 (818), 345.
20. Gruen, D.; Brimioulle, F., Selection biases in empirical p (z) methods for weak lensing. *Monthly Notices of the Royal Astronomical Society* 2017, 468 (1), 769–782.
21. Gerdes, D. W.; Sypniewski, A. J.; McKay, T. A.; Hao, J.; Weis, M. R.; Wechsler, R. H.; Busha, M. T., ArborZ: Photometric redshifts using boosted decision trees. *The Astrophysical Journal* 2010, 715 (2), 823.

22. Carrasco Kind, M.; Brunner, R. J., TPZ: Photometric redshift PDFs and ancillary information by using prediction trees and random forests. *Monthly Notices of the Royal Astronomical Society* 2013, 432 (2), 1483–1501.
23. Masters, D.; Capak, P.; Stern, D.; Ilbert, O.; Salvato, M.; Schmidt, S.; Longo, G.; Rhodes, J.; Paltani, S.; Mobasher, B., Mapping the galaxy color–redshift relation: Optimal photometric redshift calibration strategies for cosmology surveys. *The Astrophysical Journal* 2015, 813 (1), 53.
24. Schawinski, K.; Turp, M. D.; Zhang, C., Exploring galaxy evolution with generative models. *Astronomy & Astrophysics* 2018, 616, L16.
25. Lanusse, F.; Mandelbaum, R.; Ravanbakhsh, S.; Li, C.-L.; Freeman, P.; Póczos, B., Deep generative models for galaxy image simulations. *Monthly Notices of the Royal Astronomical Society* 2021, 504 (4), 5543–5555.
26. Myles, J.; Alarcon, A.; Amon, A.; Sánchez, C.; Everett, S.; DeRose, J.; McCullough, J.; Gruen, D.; Bernstein, G.; Troxel, M., Dark energy survey year 3 results: Redshift calibration of the weak lensing source galaxies. *Monthly Notices of the Royal Astronomical Society* 2021, 505 (3), 4249–4277.
27. Abbott, B. P.; Abbott, R.; Abbott, T.; Abernathy, M.; Acernese, F.; Ackley, K.; Adams, C.; Adams, T.; Addesso, P.; Adhikari, R., Observation of gravitational waves from a binary black hole merger. *Physical Review Letters* 2016, 116 (6), 061102.
28. Cuoco, E.; Powell, J.; Cavaglià, M.; Ackley, K.; Bejger, M.; Chatterjee, C.; Coughlin, M.; Coughlin, S.; Easter, P.; Essick, R., Enhancing gravitational-wave science with machine learning. *Machine Learning: Science and Technology* 2020, 2 (1), 011002.
29. Robinet, F.; Arnaud, N.; Leroy, N.; Lundgren, A.; Macleod, D.; McIver, J., Omicron: A tool to characterize transient noise in gravitational-wave detectors. *SoftwareX* 2020, 12, 100620.
30. Brownlee, J. Logistic regression for machine learning. https://machinelearningmastery.com/logistic-regression-for-machine-learning/ (accessed December 15, 2021).'

PART III

SHOWDOWN

8

AI FOR THEORY OF EVERYTHING

YANG-HUI HE AND VOLKER KNECHT

Finally, we come to an area of physics considered to involve the most comprehensive data in the theoretical sciences. The field is worthy of the amount of data implicated, as it refers to the best candidate for a theory of everything.[1–3] Hence, this area is destined to be explored with the help of machine learning. The latter is used by mapping central questions related to the theory of everything to a classical image recognition problem, as we will describe below.

Thus, machine learning is employed to estimate the amount of data itself, as it allows the number of possible geometries of space-time predicted by the theory to be assessed, as we will also illustrate below. We already see that geometry plays a central role in this theory. In fact, the link between geometry and physics has been known for a long time, and we will start with a brief history of this topic.

PHYSICS AND GEOMETRY

Johannes Kepler declared, where there is matter, there is geometry. Time and again, this vision has proven prophetic in fundamental physics. For example, Newton's theory of gravity became incorporated into Einstein's general relativity, an application of Riemann's geometry describing curved space. As Einstein lay on his deathbed

DOI: 10.1201/9781003245186-11

in 1955, he and other visionaries of his time had a dream, cherished for the last decades of his remarkable life: that there should exist a single set of equations, a single principle, describing the fundamental laws of nature. This spirit of "unification" is the inspiration for all theoretical physics. However, that Einstein's general relativity is incompatible with quantum theory became the last great hurdle.

STRING THEORY

Enter string theory,[4, 5] the brainchild of the geometrization tradition. In the 1980s, theoretical and mathematical physicists stumbled on an apparent way out. It constituted a paradigm shift in the understanding of the world. String theory proposed that the reason why there was the issue with uncancelable infinities of quantum gravity was that we had inherently assumed since Newton that elementary particles were point-like. However, it is precisely this innocent assumption which caused the untamable infinities: by allowing interactions at a single point in spacetime, energy was allowed to be concentrated at zero volume. As we recall from our childhood lessons, division by zero gives problems!

But smear the point out. Extending a point, an object of zero-dimension, gives a one-dimensional object, a line, which could be either closed into a loop or open like a segment (as shown in Figure 2.2A).[4] This line is the superstring, the fundamental constituent of everything, the Democritean *atomos* (the indivisible) in the truest sense. This single generalization cured the infinities and led to a consistent quantum theory of gravity. Indeed, all particles, forces, interactions, even spacetime itself, became different vibration modes of the string. All of reality is reduced to a cosmic string symphony, resonating harmoniously to give the rhythms of time, space, and matter.

EXTRA DIMENSIONS

But there is a problem. This unified stringy description of everything relies on supersymmetry and six extra spatial dimensions.

One might be conservative and go back to the drawing board. Or one might think a little deeper and be bolder.

The very *definition* of quantum gravity is at a tiny scale called the Planck length, which is the smallest distance measurable when considering quantum physics and general relativity at the same time. Hence, at this scale something may happen to nature which is not yet described by one of those established theories!

The Planck length is about 1.616×10^{-35} m. This is about 20 orders of magnitude smaller than the tiniest scales currently observable at CERN. The daring proposal is simply that we *do* live in ten spacetime dimensions, and the seemingly missing six dimensions are curled at this tiny length scale (as shown in Figure 2.2B) so that we only see four dimensions. Concerning the curling up of extra dimensions, an analogy would be as follows. Suppose an ant is crawling on a twig. To the ant spacetime is four-dimensional, but to a person far away, the ant seems to live in two spacetime dimensions; the relatively small thickness of the twig is too tiny to be seen. We are like the ant who thinks spacetime is only four-dimensional, while six more tiny space-like dimensions are curled up and not yet observed.

WHY STRING THEORY?

Which brings us to the common critique of the whole affair. *Nothing* is as yet observed, neither supersymmetry, nor extra tiny dimensions. While CERN, together with many leading particle accelerators of our day, is very actively looking for signatures of both, the absence of any signals for the last couple of decades has been trying the patience of many.

However, time and again, theory pre-dates any experimental data. For example, Einstein's formulation of spacetime as a Riemannian manifold was a pure mathematical conception before any evidence of gravity being the curvature of spacetime was actually measured. Furthermore, Higgs' mechanism for mass generation of particles based only on a symmetry principle had to wait decades before the Higgs boson was discovered.

Beyond the field of physics, string theory ideas of supersymmetry and the shapes of extra dimensions are engendering completely new approaches in diverse fields in pure mathematics, solving a range of concrete problems such as in enumerative geometry or in number theory.

MACHINE-LEARNING THE LANDSCAPE

We are confronted with the problem of studying the shape of six-dimensional objects. The mathematical branches which deal with these things are differential/algebraic geometry/topology.* The technical word for "shapes" is *manifolds* and we are compelled to study six-dimensional ones. The first appearance of manifolds in fundamental physics was via general relativity, when Einstein introduced this notion from Riemann's geometry to the spacetime continuum.

The difficulty with manifolds is their abundance with increasing dimension. Two-dimensional ones† are characterized by a single integer called the *genus*, denoting the "number of holes". For instance, the surface of a sphere, the simplest two-dimensional manifold, has no holes and is thus genus zero. Every other surface with no "holes" can be continuously deformed to a sphere. The surface of a donut has a single hole and is genus one. Every other surface with a single "hole", such as the surface of a coffee mug, can be continuously deformed to that of the donut. This goes on for higher and higher values of the genus – one can think of the higher genus cases as surfaces of pretzels with more and more holes. This integer "genus" thus completely characterizes the topology of two-dimensional manifolds. Moreover, this number is intimately related to the curvature of the surface, which is positive for a sphere, zero for the donut,

* The four combinations of the adjectives and nouns here constitute a set of very active and intertwining fields of study.

† In this writing, we only consider smooth, compact, orientable manifolds without boundary.

and negative for higher genus surfaces. These are classical results of Euler, Gauss, and Riemann.

Manifolds of higher dimension are much less tamed and not even classified. Even in dimension three, for example, how manifolds are organized is unwieldy. It is partially captured by the celebrated Poincaré conjecture,* which took over a century to settle, to be finally proven by the eccentric Russian genius Grigori Perleman in 2006. But we need to work in dimension six!

THE STRING LANDSCAPE AND VACUUM DEGENERACY PROBLEM

Physical constraints compel us to look not at the entire space of manifolds in dimension six but only at small corners thereof. Still, the situation is daunting. One of the standard solutions is to take the manifold to be a special type called a Calabi–Yau manifold.† Still, the known number of such manifolds is vast, as discussed below.

The bottom-line is that each six-dimensional manifold M gives a different four-dimensional universe, with its characteristic elementary particles and cosmology.‡ For each manifold M, there is a topological quantity yielding the corresponding number of quark and lepton generations (which is three in the Standard Model of particle physics, as described in Chapter 4 and shown in Figure 4.1). Overall, there is a dictionary between the topology and geometry of M and

---

* It states that a geometric object without a hole can be deformed into a sphere. And this holds not only for a two-dimensional surface in three-dimensional space, but also a three-dimensional surface in four-dimensional space.

† Named after Eugenio Calabi who conjectured the relation between the curvature and topology for so-called Kähler manifolds in the 1950s and Shing-Tung Yau who proved it in 1987 in his Fields-Medal-winning paper.

‡ This yields a multiverse – one out of nine types of multiverses discussed in Greene, B., *The hidden reality: Parallel universes and the deep laws of the cosmos*. Vintage: 2011.

the particle physics and gravity of this universe. Establishing this dictionary, performing the requisite geometric computations of M, and seeing whether the resulting universe is anything akin to ours, is the field of string phenomenology.[6] This is undoubtedly one of the most important pursuits in the theoretical sciences and a perfect realization of that old Keplerian dream of matter from geometry! This field of theoretical physics, where one uses the geometry (and more generally, algebra, arithmetic, and combinatorics) of manifolds to "create" physical worlds in terms of quantum field theories and gravity, has been affectionately dubbed *Geometrical Engineering*.[7]

How can a manifold be constructed? In dimension two, we can plot and picture them, but dimension six is quite beyond human visualization. This is where *algebraic geometry* comes in. While this may sound like a sophisticated branch of mathematics, we were actually exposed to it from our school days – we were just not told what we were doing. One might recall that $x^2 + y^2 - 1 = 0$ defines a circle, or, more technically, that the locus of the pair of points x, y in the real plane R^2 defined by the vanishing of the polynomial $P(x, y) = x^2 + y^2 - 1$ in variables x, y defines a one-dimensional shape called a circle. Voilà, we have just constructed a one-dimensional manifold! The field of algebraic geometry is about finding loci of multiple polynomials in multiple variables* and using the *algebra* of polynomial equations to study the *geometry* of the loci, such as their topology and curvature.

Now that we know how to construct manifolds, the problem is – as the astute reader might notice from the combinatorics of multivariable polynomials – that there are simply too many ways to construct them. Indeed, this reflects the fact that there is a plethora of manifolds in high dimension. In dimension six, even if we restrict to the Calabi–Yau class of manifolds, so far people have constructed billions!

---

* The fancy way of saying this is to realize manifolds as algebraic varieties.

In fact, adding further geometrical structures,* the number of possible solutions – so-called vacuum configurations – has been estimated[8] to be of the order of the often-quoted 10^{500}. Other estimates of the number of vacua are 10^{756} (via machine learning in the form of a decision tree and linear regression)[9] and even $10^{272,000}$ from a method by Ashok and Douglas.[10] This is called the *vacuum degeneracy problem* of the string landscape.

Thus, we enter the heart of the challenge. Yes, string theory is a ToE in ten dimensions of spacetime, unifying all interactions, particles, and spacetime; yes, we can construct a multitude of explicit six-dimensional manifolds; but which manifold gives a universe even remotely close to ours? How does one select *our* universe? While there has been some success in finding the exact Standard Model particle content[6] or performing statistical estimates,[6] sifting through the vast landscape computationally by brute force is simply infeasible.

MORE ON MACHINE-LEARNING THE LANDSCAPE

Thus, now more on machine learning. Confronted with the aforementioned large databases of geometrical data, constructed over the decades, it is only natural that one approaches them with techniques from our Age of Big Data. Now, could ML models with integer configurations encoding multi-degrees of manifolds as input and integers representing topological quantities as output be designed?

Yet, with a mixture of trepidation and optimism, such an experiment has been attempted,[1–3, 6] and with surprising success. A simple feed-forward neural network of only a few layers, consisting of linear and sigmoid (smooth-step-like) activation functions, was able to compute topological quantities to higher than 99 percent precision in a matter of seconds or minutes. This is in stark contrast with

* Such as considerations of vector bundles, homological cycles, etc., thereby entering the deep technicalities of algebraic geometry.

the standard computation in algebraic geometry, known (using so-called Gröbner bases) to be doubly exponential in complexity and thus computational expense. How neural networks and other classifiers and regressors are precisely doing algebraic geometry *without* any knowledge of the underlying mathematics is both puzzling and exhilarating.

Concerning the usage of ML to study the string landscape, it is worth going into a little more detail. Generalizing our discussion above, we note that a manifold in algebraic geometry is given by the intersection of K polynomials in n variables. We emphasize that the variables are not spacetime coordinates but those of the algebraic geometry used to characterize the internal Calabi–Yau manifold. So one representation would be to keep track of the list of coefficients of all the possible monomials in the variables, for each polynomial. Actually, the situation is simpler for us because we are interested in *topological quantities*, at least for the initial problem of looking for the content of Standard Model particles (without considering their masses and other non-integer parameters).* By definition, these topological quantities are independent of particular choices of the coefficients. For instance, a circle and an ellipse are topologically identical while the quadratic polynomials in two variables that define them differ in coefficients which govern the shape.

For our purposes, a Calabi–Yau manifold can be represented by an integer matrix, where each row corresponds to a respective variable and each column to one of the defining polynomials and the entries to the corresponding degrees (the highest exponent of the respective variable occurring in the corresponding polynomial), except for the first column giving the dimension of the complex

* For the next-level problem of computing particle masses and Yukawa-couplings in the Lagrangian, one needs to compute non-topological quantities such as metrics, and detailed information such as coefficients of the defining polynomials will be needed.

projective space from which the variables originate.* For instance, the most famous Calabi–Yau manifold, the so-called quintic (degree five) living inside the (four-dimensional) so-called complex projective space CP^4, is represented by the matrix [4 | 5] (corresponding to a polynomial in a single variable of degree five in the four complex dimensions of CP^4). Similarly, a Calabi–Yau manifold defined by the intersection of two polynomials of degrees (3,0,1) and (0,3,1), living inside the product $CP^2 \times CP^2 \times CP^1$ between three complex projective spaces, is represented by $\left[\begin{array}{c|cc} 2 & 3 & 0 \\ 2 & 0 & 3 \\ 1 & 1 & 1 \end{array}\right]$.

The key point is that all (families of) algebraic varieties, and thus all manifolds in particular, can be represented in this way! Next, we need to compute topological quantities, such as Betti numbers† (or their complex counterpart, the Hodge numbers), that correspond to particle content upon reducing to the Standard Model.

The study of modern geometry, notably the Bourbaki School from the mid-20th century, has precisely done this for us. It involves heavy mathematical machinery, which not only involves mathematical sophistication, but is also computationally expensive. The key bottleneck to algebraic geometry is the Gröbner basis mentioned above, which is famously doubly exponential in complexity with respect to the degree and number of variables.

Yet, at the end of the day, we are simply extracting a positive integer from a matrix. In fact, to make the problem even more familiar to data scientists, we can represent the matrix input as a pixelated

---

* The original representation, in a class of manifolds called CICYs by Candelas and colleagues, is to embed the Calabi–Yau manifold into a product of complex projective space so that the rows correspond to the degree of the corresponding factor within the product.

† The first three of them, for example, give the number of contiguous paths, two-dimensional holes, or three-dimensional voids, respectively, of a manifold. For a torus, they are one, two, or one, correspondingly.

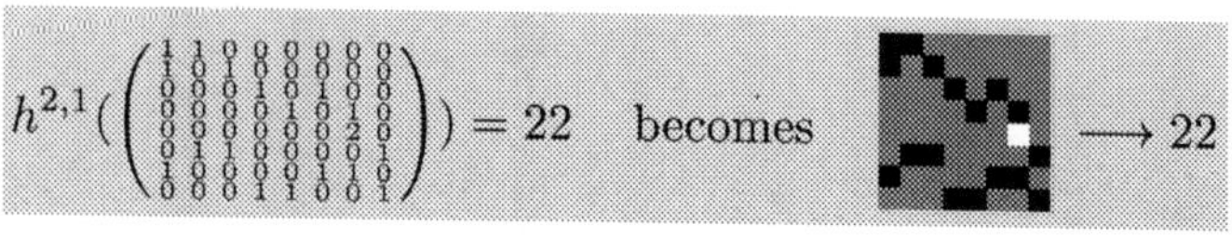

$$h^{2,1}\left(\begin{pmatrix} 1&1&0&0&0&0&0&0\\ 1&0&1&0&0&0&0&0\\ 0&0&0&1&0&1&0&0\\ 0&0&0&0&1&0&1&0\\ 0&0&0&0&0&0&2&0\\ 0&1&1&0&0&0&0&1\\ 1&0&0&0&0&1&1&0\\ 0&0&0&1&1&0&0&1 \end{pmatrix}\right) = 22 \quad \text{becomes} \quad \longrightarrow 22$$

Figure 8.1 Machine learning theory of everything. The Calabi–Yau manifold X is represented by an integer matrix, and the Hodge number $h^{2,1}$ is computed to be 22. This is then represented via color-coding the matrix, yielding a pixelated image labeled 22.

image (one might need to regularize over the data set by filling with zeros) and establish a labeled data set of images. This was the first insight of the aforementioned paper[1] by the first author.

We illustrate this idea using the following example depicted in Figure 8.1: the Calabi–Yau manifold X is an integer matrix and one of the key topological numbers, the Hodge number $h^{2,1}$, is here computed to be 22 using the traditional expensive method. This is then represented as the labeling of 22 to the pixelated image where 0 is gray, 1 is black, and 2 is white, say. All of a sudden, computations in algebraic geometry become no different in nature to a hand-written digit recognition problem.

We can then pass such data – luckily over the decades, the physics and mathematics community have been compiling such geometrical data by brute-force computation – to a standard supervised ML algorithm. It was found[1, 3] that a simple feed-forward neural network with three or so hidden layers and sigmoid activation functions, or a support vector machine, at five-fold cross-validation can already achieve an accuracy percentage in the high 90s. In a way, this is a great demonstration of the universal approximation theorems of neural networks: even a mapping as complicated as cohomology can be piece-wise approximated by deep/wide networks.

Yet, the implications are as surprising and puzzling as they are profound. What is the ML algorithm actually doing? It is certainly spotting patterns yet unseen by traditional mathematics, but how do we interpret them? Indeed, when 100 percent accuracies are reached in validation, surely this will give room for conjecture formulation in mathematics.

Since 2017, ML has become an integral part of the string theory community. A brainchild of mathematics and theoretical physics, string theory has now embraced modern data science. To unleash such contemporary methodology to address the string landscape is not only inevitable, but also necessary. Indeed, as the title of this chapter suggests, the time is ripe for "AI for Theory of Everything".

EPILOGUE

Emboldened by ML's surprising success in algebraic geometry, one cannot resist setting AI to explore the landscape of all of mathematics. Do "data" from different branches of mathematics – algebra, geometry, arithmetic, graph theory, combinatorics, etc. – have inherently different patterns or degrees of difficulty? Is there a hierarchy of complexity within mathematics itself?

So there is an increasingly important role for computers and AI in mathematics, particularly in the area of automated theorem-proving.[11] While one might call such axiomatic AI "bottom-up" mathematics, our proposed paradigm of conjecture formulation and pattern recognition using ML on available mathematical data should constitute a complementary "top-down" approach to mathematics.[12]

REFERENCES

1. He, Y.-H., Deep-learning the landscape. *arXiv preprint arXiv*:1706.02714 2017, 1–40.
2. He, Y.-H., Machine-learning the string landscape. *Physics Letters B* 2017, 774, 564–568.
3. He, Y.-H., *The Calabi–Yau landscape: From geometry, to physics, to machine learning*. Springer Nature: 2021; Vol. 2293.
4. Greene, B., *The elegant universe: Superstrings, hidden dimensions, and the quest for the ultimate theory*. American Association of Physics Teachers: 2000.

5. Yau, S.-T., *The shape of inner space: String theory and the geometry of the universe's hidden dimensions*. Basic Books: 2012.
6. Candelas, P.; Horowitz, G. T.; Strominger, A.; Witten, E., Vacuum configurations for superstrings. *Nuclear Physics B* 1985, 258, 46–74.
7. Katz, S.; Klemm, A.; Vafa, C., Geometric engineering of quantum field theories. *Nuclear Physics B* 1997, 497 (1–2), 173–195.
8. Kachru, S.; Kallosh, R.; Linde, A.; Trivedi, S. P., De Sitter vacua in string theory. *Physical Review D* 2003, 68 (4), 046005.
9. Carifio, J.; Halverson, J.; Krioukov, D.; Nelson, B. D., Machine learning in the string landscape. *Journal of High Energy Physics* 2017, 2017 (9), 1–36.
10. Ashok, S. K.; Douglas, M. R., Counting flux vacua. *Journal of High Energy Physics* 2004, 2004 (1), 060.
11. Buzzard, K. Will computers outsmart mathematicians? https://www.gresham.ac.uk/lectures-and-events/smart-computers (accessed December 30, 2021).
12. He, Y.-H., Machine-learning mathematical structures. *arXiv preprint arXiv:2101.06317* 2021.

9

CONCLUSION AND OUTLOOK

VOLKER KNECHT

Physics is the most fundamental branch of natural sciences and provides the foundations of engineering and thus all modern technology. An ever-increasing role in research in this area is played by AI and especially machine learning, which offers an increasingly indispensable tool to handle the exponentially growing amount of data involved.

AI can be realized via a wide variety of algorithms and facilitates the modeling and exploration of the physical world from subatomic to cosmic scales. Whether elementary particles, molecules, or condensed matter, forecasting earthquakes[1] or hurricanes,[2] studying black holes, galactic lenses, the cosmic web, or an advanced though controversial[3] candidate for a theory of everything involving the most extensive data in science – AI is crucial in advancing the field. Thanks to AI, great scientific challenges are mastered, allowing incredible progress otherwise impossible.

AI is becoming crucial for tasks like separating signals from overwhelming noise, classifying objects from the smallest particles to whole galaxies, enhancing and deriving new models for computer simulations, characterizing complex quantum states matter,

DOI: 10.1201/9781003245186-12

designing new materials, and exploring a putative string multiverse. Finally, it even shows potential to discover new fundamental laws.

Inversely, physics boosts AI; it provides theoretical frameworks to inspire, understand, and advance machine learning algorithms. Furthermore, it heralds a new era of information technology via quantum computing, enabling the implementation of a new class of algorithms called quantum machine learning. In November 2021, IBM presented a breakthrough quantum processor called "Eagle" cracking the 100-qubit mark with 127 qubits. That renders "Eagle" so superior that it can no longer be simulated by a conventional supercomputer: the company declares that simulating the "Eagle" would need "more classical bits [than] there are atoms in every human being on the planet".

By 2023, IBM would like to have ultimately broken through the "sound barrier" of 1,000 qubits; the planned chip called "Condor" shall reach 1,121 qubits. Experts claim this is the barrier a quantum processor must overcome to be for the first time more efficient than a conventional processor not only for exotic test cases but in a robust general sense.[4, 5]

This progress will boost AI and help to overcome current limitations in the field. These shortcomings and the largely still not exploited potential of AI become apparent when comparing artificial with human intelligence. Undoubtedly, AI is much faster than humans. AI can – for instance – read cardiac magnetic resonance imaging scans almost 200 times faster than humans, with a precision equivalent to experts.[6]

Nevertheless, in some respects the human brain is still far superior to AI. This is why it may even now serve as a template and role model to further improve AI via physics-based technical innovations. Above all, the human brain is far more energy efficient. A representative of the largest supercomputers consumes ten million watts and performs about eight billion megaflops – one megaflop being 10^6 floating point operations per second. The human brain can complete about two billion mega flops using only 20 watts.[7] This means that in terms of megaflops per watt the human brain is a million times more energy efficient than a supercomputer; thus, our brain is really high-tech!

As another example, our brain is more robust against noise on the data when it comes to image recognition. For instance, perturbing the photo of a turtle with noise barely perceived by a human can cause the AI to mistake the animal for a rifle.[8] AI can still learn a lot from us!

As indicated above, great advancements in AI come from physics which inversely profits from AI. The mutual reinforcement of physics and AI leads to exponential progress in both fields. This could lead to a point in time when AI surpasses human intelligence and would thereby rapidly improve itself and make new inventions. This point is denoted as technological singularity – or simply the singularity.[9] When will it happen?

Singularity is related to artificial general intelligence (AGI) – the hypothetical ability of an intelligent agent to understand or learn any intellectual task that a human being is able to. Four polls of AI researchers suggested a median probability estimate of 50 percent that AGI would be developed by 2040–50.[10, 11]

What does it mean for science? The famous physicist Stephen Hawking expected in 2014 it could imply that AI may out-invent human researchers.[12] Then AI would drive science forward by itself, for the sake of technology and human knowledge; the power of AI for physics would then fully unfold.

REFERENCES

1. Beroza, G. C.; Segou, M.; Mostafa Mousavi, S., Machine learning and earthquake forecasting: Next steps. *Nature communications* 2021, 12 (1), 1–3.
2. Boussioux, L.; Zeng, C.; Guénais, T.; Bertsimas, D., Hurricane forecasting: A novel multimodal machine learning framework. *Weather and Forecasting* 2022, 37.
3. Bradlyn, B. J. String theory: A controversy in ten dimensions. http://web.mit.edu/demoscience/StringTheory/index.html (accessed January 23, 2022).

4. Gerstl, S. Codename „Eagle“: IBM präsentiert Rekord-Quantenprozessor mit 127 Qubits. https://www.elektronikpraxis.vogel.de/ibm-praesentiert-rekord-quantenprozessor-mit-127-qubits-a-1075506/ (accessed January 23, 2022).
5. Hugh Collins, K. E. IBM Unveils breakthrough 127-qubit quantum processor. https://newsroom.ibm.com/2021-11-16-IBM-Unveils-Breakthrough-127-Qubit-Quantum-Processor (accessed January 23, 2022).
6. Machine learning could offer faster, more precise results for cardiac MRI scans. https://www.sciencedaily.com/releases/2019/09/190924080037.htm (accessed January 23, 2022).
7. Cooper, K. Why do computers consume more energy than the human brain? https://www.quora.com/Why-do-computers-consume-more-energy-than-the-human-brain/log (accessed January 23, 2022).
8. Hossenfelder, S., 10 Differences between artificial intelligence and human intelligence. In *Science without the goobledygook*. YouTube. 2019.
9. Cadwalladr, C., Are the robots about to rise? Google's new director of engineering thinks so. *The Guardian* 2014, 22.
10. Khatchadourian, R., The doomsday invention. *The New Yorker* 2015, 23.
11. Müller, V. C.; Bostrom, N., Future progress in artificial intelligence: A survey of expert opinion. In *Fundamental issues of artificial intelligence*. Springer: 2016; pp 555–572.
12. Hawking, S., Transcendence looks at the implications of artificial intelligence: But are we taking AI seriously enough?. https://cacm.acm.org/ (accessed January 23, 2022).

APPENDIX: TABLE OF CONTENTS FOR ELECTRONIC SUPPLEMENT

Electronic Supplementary Material is available for free download from the publisher's website.

Here is its *table of contents*:

- More on Machine Learning in Cosmology
  - Accurate AI – an Unusual Challenge
- Credits for Copyrighted Material (Graphics): Additions
- List of References

INDEX

C

D

E

F

G

H

I

N

O

P

Q

R

S

T

U